ENGINEERED ORGANISMS in ENVIRONMENTAL SETTINGS

Biotechnological and Agricultural Applications

ENGINEERED ORGANISMS in ENVIRONMENTAL SETTINGS

Biotechnological and Agricultural Applications

Edited by
Morris A. Levin
and **Eitan Israeli**

CRC Press
Boca Raton   New York   London   Tokyo

Library of Congress Cataloging-in-Publication Data

Engineered organisms in environmental settings : biotechnological and
    agricultural applications / edited by Morris Levin and Eitan Israeli.
        p.   cm.
    Includes bibliographical references and index.
    ISBN 0-8493-4465-4 (alk. paper)
    1. Transgenic organisms.   2. Biotechnology.   3. Agricultural
biotechnology.   I. Israeli, Eitan, 1944–   .   II. Levin, Morris A.
QH442.6.E54   1996
660′.6—dc20   95–40472
                                                                            CIP

Morris Levin, Editor

Dr. Levin received his B.S. in Biochemistry from the University of Chicago and his Ph.D. in Bacteriology and Biophysics from the University of Rhode Island. He has been employed by the Department of Defense and the Department of Interior as an environmental microbiologist, studying the survival and persistence of microbes in environmental situations and their effect on public health. He joined the US EPA at its inception and has continued working in the areas of public health and microbiology. He was involved in Biotechnology Risk Assessment at the EPA and has continued risk assessment research since 1989 when he joined the University of Maryland and Biotechnology Institute's Center of Public Issues in Biotechnology . He is active in international affairs, serving as chairman of the IRRO, an organization devoted to enhancing the availability of environmental release data on a global basis.

Eitan Israeli, Editor

Dr. Israeli received his M.Sc. and Ph.D. from Tel Aviv University, Department of Microbiology. His research has included problems of freeze drying and aerosols of bacteria, as well as genetic engineering of bacterial plasmids. As a visiting scientist he spent time at USC, MIT and with the EPA where he studied the problems of environmental release of genetically engineered strains. Since 1970 he has been with IIBR where he holds the positions of Head of the Department of Infectious Disease and Biosafety Officer. Since 1980 he has specialized in biohazards and safety laboratories. Dr. Israeli consults on biosafety to universities, research institutions, hospital, and industry in Israel. He lectures at Tel Aviv University and the Weizmann Institute. He is a member of the National Committee for Guidelines and Supervision of Engineered Plants, and is chairman of a National Committee on safety and hygiene.

Dr. Israeli is a member of AAAS, NYAS, AB2SA, and professional associations in Israel. His biography was published in the 1995 edition of *Who's Who in the World*, and he is an editor-in-chief of *Synthesis-On-Net*, a life science journal on the Internet. Dr. Israeli is an author of many scientific articles, including reviews and chapters in books on biosafety and published the only book in Hebrew on this subject. He has also published a science fiction novel, *The Amazon Conexion*, which deals with possible effects of environmental uses of genetically engineered products. Dr. Israeli was born in Israel in 1944, is married, and has two children.

CONTRIBUTORS

Nancy L. Brooker, B.S., M.S, Ph.D.
Assistant Professor of Biology
Department of Biology
Pittsburg State University
Pittsburg, Kansas

William Bruckart, B.S., M.S., Ph.D.
Research Plant Pathologist
ARS-FDWSRU
United States Department of Agriculture
Frederick, Maryland

Robert S. Burlage, Ph.D.
Staff Scientist
Department of Environmental Sciences
Oak Ridge National Laboratory
Oak Ridge, Tennessee

Richard Fink, B.S.
Associate Biosafety Officer
Biosafety Office
Massachusetts Institute of Technology
Cambridge, Massachusetts

David A. Fischhoff, Ph.D.
Director, Technology Strategy
Ceregen
Monsanto Corporation
St. Louis, Missouri

Cyril G. Gay, D.V.M, Ph.D.
Associate Director
Department of Regulatory Affairs
Pfizer, Inc.
Lincoln, Nebraska

Elizabeth A. Gilman, M.S.
Assistant Biosafety Officer
Biosafety Office
Massachusetts Institute of Technology
Cambridge, Massachusetts

Eitan Israeli, Ph.D.
Head
Department of Infectious
  Diseases
Israel Institute for Biological Research
Ness-Ziona, Israel

Morris Levin, Ph.D.
Massachusetts Institute of Technology
Research Professor
Center for Public Issues
University of Maryland Biotechnology
  Institute
Baltimore, Maryland

Daniel F. Liberman, B.S., M.S., Ph.D.
Director
Department of Environmental Health
  and Safety
Boston University
Boston, Massachusetts

Sally G. Metz, M.S.
Business Director
Monsanto Corporation
St. Louis, Missouri

Ilan Sela, Ph.D.
Professor of Virology and
  Molecular Biology
The Hebrew University of Jerusalem
Faculty of Agriculture
Rehovot, Israel

Lidia S. Watrud, Ph.D.
Research Ecologist
Biotechnology Program Leader
Department of Terrestrial Plant Ecology
United States Environmental Protection
  Agency
Corvallis, Oregon

TABLE OF CONTENTS

Chapter 1

INTRODUCTION AND OVERVIEW

Eitan Israeli and Morris Levin

TABLE OF CONTENTS

1   SCOPE

Biotechnology has grown from one company in the U.S. in 1971 to over 1500 today. Initially, the focus was on research, conducted at federal, university, and commercial establishments, and was primarily focused on research into potential pharmaceutical products. This has resulted in over 1000 biotechnology products in commerce in the pharmaceutical sector in the U.S. The same trend is found in the development of biotechnology in other parts of the world. Biotechnology has been identified as a major factor in the economic growth of Japan and is being focused on for development of both science and industry. Europe has formed a joint program (Biotechnology Action Programe[7]) within the European Commission.

The first biotechnology product was a pharmaceutical (insulin, in the U.S.). As the industrial potential of the science became more evident, interest shifted to other industrial sectors. Agricultural, waste treatment, and pollution control methodologies became subjects of molecular biology researchers. The first agricultural Agbiotech product was related to the pharmaceutical industry, a swine vaccine marketed by Syntro Corporation. Agricultural, pollution control, and waste treatment products all involve field applications — release of the biotechnology product to the environment.[29]

Biological processes have traditionally been applied in all of these fields. However, enhanced interest as a result of the advent of biotechnology resulted in improvement of existing techniques, increased research to improve both the processes and the microbes that were being utilized, and the end product. Prior to the work of Cohen et al.,[5] transgenic microorganisms were nonexistent. Modifications to microorganisms to increase the range of compounds degraded and the degradation rate ability were achieved by classical methods (e.g., transformation, UV mutagenesis). With the application of the tools of molecular biology, it is possible to rapidly develop specialized organisms tailored for particular uses. To date, these organisms have been developed for use in agriculture, pollution control, and waste treatment. They have been released to the environment in agricultural situations. Although waste treatment and pollution control have been active areas of research and development, no releases of engineered organisms have occurred.[14] This topic is developed in Chapter 4.

Agricultural releases have become commonplace. Over 2000 releases have occurred, depending on how one defines "release".[1,3,8,9,22,28] Given that source data are so variable, there is little value in attempting to derive a specific number for releases around the world or in a particular country. Each country — and in many countries agencies within a given government — has its own definition of release. (In some instances the phrase environmental application is substituted for release.) It is valuable to focus on trends. Beck and Ulrich[1] point out that the overall number of transgenic plant tests is increasing rapidly. This is supported by reports from the U.S. Department of Agriculture (USDA), U.S. Environmental Protection Agency (USEPA), *GeneExchange*, and Organization for Economic Cooperation and Development (OECD). No two sources agree as to exact numbers, but it is clear that from 1988 to 1992 the number of transgenic plant tests approximately doubled each year, starting from 33 in 1988 to approximately 400 in 1992. Interestingly, there has been a change in the relative frequency of tests by country. The ratio of tests (North America/Europe) changed from approximately 1.2 to 4.8 over the same period, indicating that more projects are reaching the field stage in North America than elsewhere. Many of the tests were conducted in Canada. The ratio of U.S./Canadian trials changed over the same period, due primarily to Canadian interest in Rapeseed.

Between 1989 and 1991 there were 1.4 trials in the U.S. for each trial in Canada. In 1992 the ratio was 0.7, with Canada conducting 189 tests to 131 in the U.S. However, the total number of tests (engineered organisms) in the U.S. continues to climb, with the USDA reporting a 92% increase from August 1993 to August 1994 (1041 permits in 1993 vs. 2037 in 1994).

It must be emphasized that these numbers are for field test permits. More than one site may be involved per permit. In the U.S. the ratio of permits to sites has changed radically. In 1988, each permit meant one site. By 1993, 4.6 sites were involved for each permit. Also, many organisms which formerly required permits are no longer regulated. Beck and Ulrich[1] state that in 1993 there were 276 such cases.

2 DEFINITIONS AND ISSUES

There are major issues which affect the ability to summarize release date:

1. Release and scope (i.e., what is included in the definition of a genetically engineered organism) are defined differently by the researcher and/or the governmental reviewing agency.

2. The type (i.e., what organism and for what purpose) as well as the scale (i.e., small-scale field trial or commercial application) of the release greatly affect the potential for beneficial or adverse effects.

3. A significant regulatory issue arises from the distinction between releases conducted for academic or commercial research purposes, which produces a recording issue.

A major problem is that "release" is defined differently in different countries and even in jurisdictions within countries. The OECD encountered this difficulty when preparing their BIOTRAC program which attempted to list all releases, on a global scale, and thus provide a database for use by scientists and regulators. Ultimately, the OECD used the definitions of the source of the information. This resulted in some entries which pertained to very small areas and single time (i.e., one small field test at a location) while other entries pertained to many sites and times (i.e., many small trials with the same organism at different times and places, generally but not always near one another).

Articles in scientific journals and other publications define release at the discretion of the author. Thus, one company which has developed an improved rhizobium will announce a release of an engineered microorganism in a number of widely separated test plots while another may announce several releases, all of the same organism, in a number of locations, all within a short distance of one another. Other, more limited databases (e.g., USDA[27]) include only requests for permits for interstate shipment approved by and therefore defined by that organization, resulting in more complete and uniform entries. Thus, the USDA database contains items ranging from small, single organism tests to multisite tests with the same organism at different times, each clearly noted. However, USDA records only U.S. trials subject to USDA regulatory authority, and it is also important to note that as time passes, agency definitions change.

The scope issue — definition of a recombinant product — has presented problems for biotechnology watchers and regulators since the industry began. The primary problem is deciding when a sufficient modification has occurred by an appropriate method to warrant the label genetically engineered organism (GEO). Many countries have new rules or modifications of existing rules to deal with GEOs. However, deciding when an organism fits the definition has always been a problem. Thus, for example, microbes modified by UV mutagenesis or plants developed as a result of selective breeding over many generations are usually not included in the GEO category, despite the fact that engineered organisms, modified by molecular biology techniques involving recombinant DNA, exhibiting identical characteristics are included. Changes in scope result in changes in recording. The first field trial of an engineered (by the 1983 scope definition) would not be considered engineered today by the reviewing agency (USEPA) and therefore the test would not be recorded.

Another aspect of recording is the size of the release and the purpose. Releases vary in purpose, and this affects the scale and overall experimental design. To date, the only large-scale, commercial use of an engineered plant or viable microbe has been in China.[4] Little is known about the specifics, but within China engineered tobacco and tomato have been planted on a commercial scale. Field tests with plants producing coat protein virus for resistance to viral diseases and *Bacillus thuringiensis* toxin to enhance resistance to insect pests have been conducted. Products have been distributed commercially. Recombinant bacteria with a higher efficiency of nitrogen fixation have been tested and commercially distributed.

Within the U.S., one product, a transgenic *Pseudomonas* to which a toxin gene from *Bacillus thuringiensis* has been added, is being commercially distributed. The product is sterilized before marketing. Transgenic plant products (tomatoes, cotton) have been approved and tomatoes marketed in 1994.

3   RISK ASSESSMENT

Development of agricultural of biotechnology products has resulted in the development of U.S. Risk Assessment protocols. The first field test of an engineered organism was for a bacterium which would protect plants against frost. The successful test was conducted in 1986.[13] Since that time over 2000 field tests of plants and microorganisms have been conducted in the U.S.[9] Most of the tests have been conducted by academic researchers seeking a better understanding of a particular organism or process or by a company with commercial interests in mind. The objective of most field tests is to demonstrate efficacy. Very few have been specifically designed to demonstrate the safety aspects of the potential product.[23] In most cases the test is designed to demonstrate that a product is efficacious and hence commercially valuable. Field tests have ranged in size form 0.25 to 10 acres. They may be repeated at different locations to test the effect of climatic change.

Much effort has gone into devising protocols for assessing the risks associated with field trials. Table 1 identifies the concerns generally evaluated when attempting to evaluate the safety aspect of a particular trial. Most developed countries and jurisdictions within countries require evaluation of data from these categories. These were codified in the U.S. as Points to Consider.[16] UNIDO has drafted a voluntary code for underdeveloped countries to use that takes into consideration science requirements for credible risk assessment and the resources of underdeveloped countries. The UNIDO document was prepared by an expert group familiar with risk assessment needs and country capabilities.

Requirements for risk assessment of large-scale, commercial use of transgenic organisms are less well defined. It cannot be assumed that the data requirements for field tests will suffice for commercial-scale releases or that the data will be applicable. A recent international symposium on large-scale use of transgenic plants[2] examined the data needs and available methods for considering commercial-scale release relative to impact on diversity, biogeochemical cycles, community structure, pollen spread, etc. Although it was generally felt that engineered plants are generally not to be considered major hazards, the need to develop specific tests and identify end points was noted.

4 REGULATORY ISSUES

Most developed countries have either drafted legislation to deal with the problem of reviewing GEOs or have adapted existing legislation through the regulatory process to achieve the same result. These have been reviewed[6,11,12,26] and a comprehensive legal review is beyond the scope of this book. The science base — that is, the scope of the information required of the applicant as a basis for review — is remarkably similar. The areas identified in Table 1 are found in all data requirement listings in one form or another. They are found the U.S. National Institutes of Health Recombinant Advisory Committee guidelines and the USDA application form

TABLE 1

Components of Points to Consider

1. Summary of Trial
        Objective
        Feasibility
        Benefits and Risks
        Justification

2. Description of Trial
        Conditions
        Location
        Site Characteristics
            probability of dissemination
            description of target and nontarget population present
        Containment
        Monitoring Procedures
        Mitigation Procedures
        General Procedures
            transportation procedures
            employee training
            security procedures
        Disinfection Procedures

3. Genetic Characteristics of Organisms (Parent and Recipient)
        Identification
            taxonomic description
        Genotype
            characterization of genetic material
        Potential for Gene Transfer
            capability and mechanism
            evidence for exchange in nature

4. Phenotype
        rationale for selection
        anticipated changes in host
        culture requirements, life cycle, habitat
        pathogenicity data
        antibiotic resistance and production
        survival and persistance data
        control mechanisms
            natural agents
            effective disinfectants

5. Introduction of Genetic Material
        How Modified
            source and function of inserted DNA
            methods used to identify, isolate, and insert DNA
        Vector
            identification
            site of gene insertion
            method of introduction to host
            characterization of inserted genes
            location, amount, stability of vector DNA

TABLE 1 (Continued)

Components of Points to Consider

---

6. Comparison of Recipient to Parent
   laboratory data describing relative survival
      persistence
      multiplication
      dissemination
7. Identification of Specific Potential Adverse Effects

---

and are used by the USEPA as a basis for decision making, as well as in Germany, Japan, Australia, and Canada. A major difference in some processes (e.g., Canadian and Australian) is the stress placed on degree of familiarity with the host organism. This is similar to suggestions put forth by the U.S. National Academy of Sciences and the Ecological Society of America.[24]

The President's Council on Competitiveness[15a] (under President Bush in 1991) released a report which raised perturbing questions concerning policy issues in this area. According to the Council, GEOs "shall not be subject to federal oversight" unless there is substantial evidence to indicate "unreasonable risks" (defined as where "full social and environmental costs exceed the cost of governmental intervention to redress it"). The report also proposes that agencies consider "social needs" when evaluating the risks of planned introductions. This report created controversy.

5 PUBLIC CONCERNS

Public reaction to the widespread use of GEOs in environmental situations, in the U.S., Germany, and Holland, for example, has been negative and hostile. Plants have been uprooted, picketers have appeared, and demonstrations have occurred. With the passing of time, as releases became more common and no adverse effects of great magnitude occurred, public opinion became more moderate. Nevertheless, there is still strong reaction and resistance to the inclusion of foreign genes in crop plants.[30] It is essential that risk assessments be conducted in a scientifically defensible manner. However, in order to achieve public acceptance of the product or the field release, the public must be convinced of the credibility of the assessment and must feel that there has been input from the public at large in the process.[26] The assessor must take into consideration how the risk is perceived by those at risk in order to ask for data appropriate to produce convincing arguments that the product/trial is safe.

Public perception of the risks and benefits of biotechnology can play an important role in the industry's attempts to test and market products.

Despite the growing number of tests, the general lack of basic scientific understanding with respect to environmental introductions has fueled the existing feelings of uneasiness with which the public views high technology. High technology — in this case, big business allied with molecular biology — has become suspect, guilty of the potential for adverse health and environmental effects and of being controlled by short-term interests guided solely by motives for immediate profit. This perception is the outcome of public encounters with other high-tech industries which began with much hype describing benefits and little attention to possible adverse effects (e.g., chemical industry with its Love Canal and Valley of Drums, nuclear industry with Chernobyl and Three Mile Island).

Public attitude toward high-tech and toward biotechnology in particular is complex and contains contradictions. In 1987 the Office of Technology Assessment (OTA) commissioned Louis Harris and Associates to conduct a survey of American public opinion toward biotechnology. The results illustrated the complexities, contradictions, and need for greater understanding on the part of the public about biotechnology and on the part of the industry and regulatory community about public concerns.

The poll revealed that only 19% of the respondents had knowledge of the potential risk posed by products of biotechnology. A much higher percentage of the population (52%) believed it at least "somewhat likely" that these products would present some serious danger to humans or the environment. In spite of this, a clear majority (66%) thought that genetic engineering would bring changes that will improve the quality of life. Respondents were much more positively inclined toward genetic manipulation of plants, animals, and microbes than toward human genetic manipulation. The conduct of small-scale field tests was acceptable to a large majority (82%) of the respondents, while only 53% felt that large-scale tests should be permitted; 55% were willing to accept environmental risk in exchange for potential benefits, even if the risk of losing some local species of fish or plant was as high as 1 in 1000. While the level of acceptance increased as the risk decreased, a majority of respondents indicated unwillingness to approve a test if the risk level were known.

The OTA was not the first organization to survey public opinion concerning biotechnology. Jon Miller at Northern Illinois University surveyed 673 science policy specialists in 1983, finding that most were not informed about the risks, implications, or the science. These are not surprising findings for 1983, since there had been no field trials and the research efforts were in the beginning phases. There have been 22 published surveys in different parts of the world since that time[30] and many others by individual companies or industries interested in specific products or research areas. For example, the World Bank funded a survey of biotechnology activity in the marine area which included a section dealing with public issues.

Of the 22 published surveys, 7 were instituted by industry, 8 by governmental groups or agencies, 3 by universities, and 4 by public organizations. It is difficult to compare results from surveys. The questions asked may appear to be the same, but slight differences in wording and subtleties in meaning will affect the response. Additionally, the means of selection of respondents may vary. The composition of respondents can materially affect the results of a survey.

A few trends can be tentatively identified. The amount of knowledge about biotechnology has increased. This is not surprising since knowledge is broadly defined as having read of heard about biotechnology. Within the U.S., knowledge has risen from 16% of the respondents to 38%. In-depth knowledge — ability to explain biotechnology or to describe DNA — is low at present and was not measured in earlier polls. The source of information for most respondents was television, either news or special programs.

Understanding responses to the question of the overall value of bio-technology is difficult. In the early polls[21] 52% felt there was risk involved, but 66% felt there would be overall benefits, and 55% of those felt that they would accept the risk for the benefits. A recent U.S. survey found that 94% of respondents felt that the benefits were worth the risk. However, the survey was of persons in the agricultural field holding advanced degrees. If one compares general surveys done in different parts of the globe (U.S., Europe) one finds a general lessening of acceptance of biotechnology. In 1991 (U.S.) 66% of respondents felt the technology would be beneficial. European data indicate that in 1990 63%, in 1991 54%, and in 1993 only 48% saw overall benefits.

These findings are supported by the growing lack of confidence in public authorities relative to safety of biotechnology — especially in the U.S., where respondents indicated that their greatest trust was in the opinions of nutritionists. Nevertheless 85% of Europeans surveyed were in favor of stronger governmental controls. Hogal and Kendall[10] conducted a wide-ranging survey of consumer attitudes in the U.S. toward biotechnology. He found that public acceptance of biotechnology was still strong and that nutritionists were the most acceptable source of information.

There is some urgency in dealing with these concerns since engineered organisms are likely to be offered for sale before the end of the decade. There appears to be a positive trend in that additional knowledge lessens resistance and fear; 88% of high school teachers, for example, feel that benefits greatly outweigh the risks.

This book is intended to provide a global summary of environmental applications of genetically engineered organisms. The objective is to categorize and describe the environmental releases and their effects. An overview of testing around the globe is presented. Tests involving microorganisms (bacteria, fungi, and viruses) are examined for efficacy and risk. Although all chapters on microorganisms deal with research issues, the virus chapter (Chapter 6) focuses heavily on research involving transgenic viruses and plants, describing possible outcomes in terms of future products and their potential benefits and risks. Research which will lead to further testing is described. Safety issues are identified and measures to assure safety are described. The current regulatory posture for reviewing field trials is presented.

REFERENCES

1. Beck, C.I. and T.H. Ulrich. 1993. Environmental release permits. *Bio/technology* 11: 1524–1529.
2. Burke, T., R. Seidler, and H. Smith. Ecological implications of transgenic plants. *Mol. Ecol.* 3(1): 1–89.
3. Chasseray, E. and J. Deusing. 1993. *Field Trials of Transgenic Plants*. AGRO Food Ind. Basel, Switzerland, 1–10.
4. Chen, Z.L. 1992. Field releases of recombinant bacterial and transgenic plants in China. In *Biosafety Results of Field Tests of Genetically Modified Plants and Microorganisms*. Biologische Bundesanstalt für Land und Forstwirtschaft, Braunschweig, Germany, 53–54.
5. Cohen, S.N., A.C.Y. Chang, H.W. Boyer, and R.B. Helling. 1975. Construction of biologically functional bacterial plasmids *in vivo*. *Proc. Natl. Acad. Sci. U.S.A.* 70: 3240–3244.
6. Cordle, M.K., J.H. Payne, and A.L. Young. 1991. Regulation and oversight of biotechnological applications for agriculture and forestry. In *Assessing Ecological Risks of Biotechnology*, L. Ginzberg, Ed. Butterworth, Boston, 289–311.
7. Economidis, I. 1990. Biotechnology R&D in the EC. Commission of the European Community, Brussels.
8. Mellon, M. and Rissler, J. 1994a. Experimental releases of genetically engineered organisms. *GeneExchange* 4(4): 12. Union of Conc. Sci., Washington, D.C.
9. Mellon, M. and Rissler, J. 1995b. Experimental releases of genetically engineered organisms. *GeneExchange* 5(2): 12. Union of Conc. Sci., Washington, D.C.
10. Hogan, T.J. and P.A. Kendall. 1992. Consumer attitudes about the Use of Biotechnology in Agriculture and Food Production. North Carolina State University, Raleigh, NC.
11. Holla, R.A.H.G. 1991. Ecological risk assessment and European Community biotechnology regulation. In *Assessing Ecological Risks of Biotechnology*, L. Ginzberg, Ed. Butterworth, Boston, 313–323.
12. Levin, M. and M. Rogul. 1991. Regulation of biotechnology by the USEPA. In *Assessing Ecological Risks of Biotechnology*, L. Ginzberg, Ed. Butterworth, Boston, 223–264.
13. Levin, M.A. and Strauss, H. 1993. Overview of risk assessment. In *Risk Assessment in Genetic Engineering*, M.A. Levin and H. Strauss, Eds. McGraw-Hill, New York, 1–17.
14. Levin, M.A. and M. Gealt. 1993. *Biotreatment of Industrial Wastes*. McGraw-Hill, New York.
15. National Institutes of Health. 1986. Guidelines for research involving recombinant DNA. *Fed. Regist.* 51: 16958–16985.
15a. President's Council on Competitiveness. 1991. *Report on National Biotechnology Policy*. Government Printing Office, Washington, D.C.

16. National Institutes of Health. 1985. Points to Consider for Environmental Testing of Recombinant DNA Microorganisms. Rec. Adv. Comm. 1985. NIH, Bethesda, MD.

17. National Academy of Sciences (NAS). 1987. *Introduction of Recombinant DNA Engineered Organisms into the Environment: Key Issues.* National Academy Press, Washington, D.C.

18. National Academy of Sciences (NAS). 1989. *Field Testing Genetically Modified Organisms: Framework for Decisions.* National Academy Press, Washington, D.C.

19. Office of Economic Cooperation and Development (OECD). 1992. Biotrac. OECD, Paris.

20. Office of Economic Cooperation and Development (OECD). 1993. Field release of transgenic plants. OECD Observer 10–12. OECD, Paris.

21. Office of Technology Assessment (OTA). 1987. Public Perception of Biotechnology. OTA, Washington, D.C.

22. Anon. 1993. Field release of modified plants. *PIP Newsletter,* September.

23. Possee, R.D. and D.H.L. Bishop. 1992. Safety tests with genetically engineered baculovirus pesticides. In *Biosafety Results of Field Tests of Genetically Modified Plants and Microorganisms.* Biologische Bundesanstalt für und Forstwirtschaft, Braunschweig, Germany.

24. Tiedge, J.M., R.K. Colwell, Y.L. Grossman, R.E. Hodson, R.E. Lenski, R.N. Mack, and P.J. Regal. 1989. The planned introduction of genetically engineered organisms: ecological considerations and recommendations. *Ecology* 70(2): 298–315.

25. UNIDO. 1989. Voluntary Code for Assessing Risks of Genetically Engineered Organisms. UNIDO, Vienna.

26. U.S. General Accounting Office. 1988. Biotechnology: Managing the Risks of Field Testing Genetically Engineered Organisms. GAO/RCED-88-27. USGAO, Washington, D.C.

27. U.S. Department of Agriculture (USDA)/APHIS Permits. 1991. Users's Guide for Introducing Genetically Engineered Plants and Microorganisms. USDA Tech. Bull. 1783. USDA, Washington, D.C.

28. U.S. Environmental Protection Agency. 1994. Microbial EUPs, Notifications and Transgenic Plant Pesticides. USEPA, Washington, D.C.

29. Witt, S.C. 1990. Biotechnology, Microbes and the Environment. Center for Science Information, San Francisco, CA.

30. Zehendorf, G. 1994. What the public thinks about biotechnology. *Bio/technology,* 12: 870–873.

Chapter 2

GENERAL OVERVIEW OF RELEASES TO DATE

Morris Levin and Eitan Israeli

TABLE OF CONTENTS

0-8493-4465-4/96/$0.00+$.50

1 ENGINEERED BIOLOGICAL CONTROL AGENTS

1.1 Introduction

Purposeful introductions of organisms are part of the classical biological control procedure and have been remarkably successful in controlling numerous weeds and pests, thereby benefitting agriculture, public health, and natural ecosystems. With a better understanding of the ecosystem, one can find native controls for alien and native pests. However, biological control agents can also damage the environment. In fact, they have been strongly implicated in the extinctions of nearly 100 species of animals worldwide. The clearest examples, and some of the best-documented extinctions known, are from islands. A few of these were the targets of the biological control agents, but most were desirable nontarget organisms. Most of the environmental impacts either were recognized circumstantially in hindsight or were discovered serendipitously by researchers studying the affected organism in the field at the critical time. Thus, the majority of the environmental impacts of biological control, including most species extinctions, undoubtedly have never been either recognized or recorded. The greater the number of introduced organisms in the environment is, the greater the potential for harm.[58] An accurate predictive theory is needed so that only the most promising and least risky alien organisms, including engineered ones, are introduced for a given purpose.

Biotechnology promises to control pests without some of the detrimental side effects of other control methods if the political and economic pitfalls that have plagued both chemical and biological control can be avoided. Use of new genetic tools may enable researchers to modify native species to control alien as well as native pests. Releasing modified native species will usually have more predictable results than releasing an alien species, although some risk will always remain.[69]

In reality, there is no easy solution to pest control. Any action to limit or kill a species will affect other species and will pose some environmental risk. Given the high reproductive potential and genetic plasticity of the insect population, the development of resistance to artificial population controls is to be expected. Some pest species have evolved ways to cope with any single control method applied in an attempt to drastically reduce their populations, whether the control is chemical, biological, cultural, or genetic. Human agroecosystems are young and are maintained for high harvestable productivity. The large acreages planted with one or a few crops amount to a multitude of bait stations. Eventually one or more species will break through the defenses and become a pest. Human's relationship to the environment, especially as related to land use and agriculture, is in a period of change. The long-term goals should be to optimize yields on a sustainable basis rather than maximize short-term returns by using a full range of control methods based on a firm knowledge of ecology and systematics. Environmental concerns can no longer be regarded as trivial, but are becoming paramount in the development of a sustainable economy.

Improvement of biological pesticides through genetic modification has enormous potential. The insect baculoviruses are particularly amenable to this approach and serve as an excellent example. Higher organisms are providing a source of genetic material to be transferred into lower organisms for pest control. A key aim of genetic engineering is to increase their speed of kill, primarily by the incorporation of genes which encode arthropod or bacterially derived insect-selective toxins, insect hormones, or enzymes. Bishop's group reported the first field trial of a genetically improved nuclear polyhedrosis virus of the alpha looper, *Autographa californica*, (AaNPV) that expresses an insect-selective toxin gene (AaHIT) derived from the venom of the scorpion *Androctous australis*.[23] Previous laboratory assay with the cabbage looper, *Trichoplusia ni*, demonstrated a 25% reduction in time to death compared to the wild-type virus, but unaltered pathogenicity and host range. In the field, the modified baculovirus killed faster, resulting in reduced crop damage, and it appeared to reduce the secondary cycle of infection compared to the wild-type virus.

The risk to the environment resulting from releases of genetically engineered organisms will be analogous to the risk from releases of classical biological control agents because many of the same ecological, cultural, political, and economic pressures will be present and also because many releases will target similar pest control problems.[58,69] Four issues have been identified as being generally applicable to the development, use, and release of transgenic organisms. These include the morality of making the organisms, possible dangers to human health resulting from the organisms, possible environmental dangers, and the acceptability of transgenic animals as food. Maclean[48] summarizes current thinking of many researchers, pointing out that the morality issue is not significant because production of transgenics is only an extension of evolution or animal breeding. He observes that most countries have regulations which govern the production of transgenics, including oversight as to the type of animal produced. Thus, environmental and health safety is assured by the various regulatory procedures. It must be remembered that all of the procedures are in effect filter systems and may allow an undesirable product to pass through. Acceptability of transgenic food products is a perception issue, where the public will decide if eating a transgenic product is worth the added cost or if the change in flavor is discernable enough to create a change in eating habits. It must be assumed that the appropriate regulatory authorities have favorably passed on the dietary safety of the product. There is much information about effects of releases — engineered and natural. Several "lessons" are clear.

1. Negative environmental effects may result from the release of novel genetically modified organisms (GMO) or natural organisms.

2. To adequately document the negative impacts, appropriate questions must be asked and trials designed to answer them.

3. Self-dispersing organisms will find their way to suitable available habitats and will not stay within the prescribed area without management.

4. Self-reproducing organisms and certain long-lived ones may affect ecosystems at many levels and in direct and indirect ways far into the future. Therefore, engineered organisms should be created so that they cannot establish wild populations, or if establishment is desired, then extra precautions concerning the risk are necessary.

5. Extinction of nontarget species is probable, but by recognizing vulnerable species early, many of these extinctions may be prevented. Therefore, vulnerable species should be monitored.

6. The use or dispersal of organisms for their own short-term benefit will be attempted. Therefore, extra precautions, security, and enforcement will be needed.

2   ENGINEERED INDUSTRIAL MICROORGANISMS

When GMOs went commercial in the 1980s the prospect of 10^{17} or more recombinant *Escherichia coli* K-12 being inadvertently released into the environment during some type of industrial accident ignited interest in the ability of *E. coli* K-12 to survive in the environment and to transfer its recombinant DNA to members of the indigenous microbial community. While previous laboratory studies indicated that the answer to both of those questions was NO, there was concern within the regulatory agencies: are commercially relevant recombinant *E. coli* K-12 strains grown to high cell densities in fermenters different from the laboratory strains? Regulatory agencies needed to be assured that commercially produced recombinant *E. coli* K-12 strains were indeed similar to the laboratory strains studied in the 1970s. However, industry was unable to provide data showing that the commercial *E. coli* strains were unable to survive or were unable to transfer their DNA to indigenous microbes. It was critical for both industry and the regulatory agencies to have these data peer reviewed so that the fate of commercially relevant strains of *E. coli* could be documented.

A special issue of the *Journal of Industrial Microbiology* (Vol. 11, 1993) contained a compilation of papers presented at the 1992 National Meeting of the Society for Industrial Microbiology in two symposia entitled "Environmental Assessment of Recombinant DNA Fermentations". Data in that issue coupled with the extensive database on laboratory strains of *E. coli* confirm that industrial fermentations with recombinant *E. coli* K-12 strains do not present environmental hazards. The studies included both *in vivo* and *in vitro* systems.

The fate of recombinant industrial strains of *E. coli* in environmental microorganisms *in vitro* was addressed in articles by Bogosian et al.[8] and Heitkamp et al.[39] In the first instance, *E. coli* strain W3100G containing pBGH1, a pBR322-based plasmid carrying the gene for bovine somatotropin (BST), was examined in Missouri river water. The authors used the highly sensitive polymerase chain reaction to look for gene transfer from the recombinant *E. coli* to indigenous microbes. Even with this technique no evidence of transfer was found.

Heitkamp et al. describe the fate of a spontaneous nalidixic acid-resistant derivate of W3110G transformed with pBGH1 in semicontinuous activated sludge units. These microorganisms were inoculated into sewage from a commercial sewage treatment plant in the St. Louis area. The recombinant *E. coli* K-12 strain could not establish itself in the microbial sewage community despite attempts to favor such an event.

Muth and Yancey studied *in vivo* survival of engineered *E. coli*. Muth et al.[54] demonstrated that their tetracycline-resistant host vector used for the production of bovine somatotropin in *E. coli* was unable to survive in male or female Fischer-344 rats even when the antibiotic tetracycline was included in the feed.

Similarly, Yancey et al.[78] found that their *E. coli* K-12 host vector (resistant to streptomycin) was unable to colonize the intestinal tracts in conventional antibiotic-treated mice. These investigators took their studies further and looked for the transfer of genetic material from the recombinant *E. coli* to indigenous microorganisms. Two *E. coli* strains were isolated from the feces of the mice, marked by isolating a spontaneous mutant resistant to streptomycin, and reintroduced into the mice. The mice were fed streptomycin to assist the colonization of the intestinal tract by the recombinant strains. The specific objective of the gene transfer experiment was to determine whether the indigenous streptomycin-resistant *E. coli* could pick up plasmid DNA (ampicillin resistance) or chromosomal DNA (tetracycline resistance) from the added recombinant *E. coli*. There was no evidence of gene transfer in any of these studies using hybridization assays.

All of these studies provide overwhelming support for the hypothesis that commercial *E. coli* strains are not intrinsically different from the laboratory K-12 strains. Furthermore, growing these strains to high cell densities does not increase the likelihood of colonization outside of the laboratory. The absence of gene transfer in all of these microorganisms indicates that there is a lack of environmental danger. While it is impossible to prove a negative, the studies reviewed here and elsewhere,[7] using state-of-the-art techniques, detected no gene transfer from recombinant *E. coli* K-12 strains to indigenous microorganisms.

3   ENGINEERED AGRICULTURAL ORGANISMS

3.1   Plants

Examples of gene flow from crops to wild relatives, producing new or more aggressive weeds, are numerous (Table 1). Whether or not this will happen depends on several factors, including the selective advantage conferred by the transgene and the relative hardiness of the first-generation hybrid in comparison to the wild relatives. If the crop-weed hybrid survives in competition with the wild relatives, then crop genes have a chance of spreading into wild populations.

Some might expect that crop-weed hybrids would not compete well against wild plants in natural settings. After all, wild plants are continually undergoing selection for the features which aid survival in those environments. But that is not necessarily the case. Instances are known where weeds have developed as a result of gene flow from nonengineered crops to wild plants (Table 1), indicating that the hybrids must have survived well enough to establish crop genes in population of wild relatives. But little experimental work has been done on the issue. Most research in the past has focused on the contaminating effect of genes flowing from the wild populations into the crop rather than the other way around.

At a conference in 1993 extensive discussions concerning potential effects on agriculture and silviculture of transgenic plants were conducted.[64] Sources of data for risk assessment based on nontransgenic plants were described. For example, pollen flow was discussed based primarily on studies of cotton crops in five states spanning the southern U.S. (Arizona to North Carolina); it was concluded that pollen flow can be described in a statistical manner and that the data are well behaved. It can be concluded that pollen will escape from the field (in a statistically predictable manner)

TABLE 1

Examples of Gene Flow from Crops to Wild Relatives Producing New or Worse Weeds

Genome flow		
From	**To**	**Result**
Crop	*Wild relative*	
Pearl millet	Wild millet	Shibra, a weed[15]
Sorghum	Johnsongrass	More aggressive types of johnsongrass[5]
Corn	Teosinte	Weedy types of teosinte[29,74]
Rice	Wild rice	Weedy rice[6]
Foxtail millet	Wild green foxtail	Weedy giant green foxtail[27,70]

and that while isolation is possible for a small field trial it is impractical on a commercial scale. Data from experiments with clover support this conclusion.[43]

For annual plants, the viability of dormant seeds presents another route of dissemination of genetic information. This could occur in agricultural and urban sites. In general, crosses between crop plants and wild hybrids have high dormancy; thus there may be the greatest potential for adverse effects in successional habitats.

Although it is clear that parameter estimates obtained from modeling pollen flow can be useful in predicting the flow distances, researchers conclude[43] that this should be a low-priority research area since absolute containment cannot be guaranteed, especially on a commercial level. The possibility of biological containment was considered. This approach offers the most potential for restricting the possibility of cross-fertilization.[46] However, given the high probability of escape, developing methods to describe/determine invasiveness potential is a critical research priority.[43]

Much information is available about naturally occurring interspecific and intergeneric hybrids. It has been pointed out that because the source of the genetic material may be only distantly related (or, perhaps, totally unrelated: i.e., from a different kingdom), the effect of transfer on the phenotype of related crops or wild plants may be greater than anticipated. Data from more closely related hybrids may be useful in predicting the outcome of more distance crosses. Although no direct data on transgenic plants are available, available data using "natural" plants could provide a basis for examining these issues. For example, in-depth studies of hybrids of oilseed rape and wild relatives, some backcrosses of which produced seed, confirm the potential of gene flow from crops to weeds producing hybrids. Adjacent crops and wild plants in the immediate area will be pollinated and the pollen will contain genetic material engineered into the crop.

Hybrids have played a significant role in crop evolution. Introgression has been known to occur in agriculture, but no evidence exists of adverse effects due to genetic exchange between crops and related species.[46]

Interest in engineered plants began in the 1980s. It was suggested at that point that experiments be designed to detect the effect and fate of the hybrids which might be produced.[64] These data are still needed. In terms of fate of experiments, existing marker genes could be used and/or transgene constructs employing a signal protein such as GUS could be used. Focused experiments in a few key species could provide data to use in decision making. Monitoring fields adjacent to transformed plants to collect data on hybridization and subsequent regrowth of hybrids would also provide a basis for decision making. This approach seems suitable for low-risk transformants and would permit moving ahead with little risk while accumulating data for decisions about other engineered systems.

Two University of California scientists, Norman Ellstrand and Terrie Klinger, recently performed careful experiments demonstrating that under controlled conditions the first generation hybrid between a wild and a cultivated radish is hardier than the wild radish.[45] In fact, the hybrids produced 15% more fruit and seeds under the experimental conditions than wild radishes.

While this advantage would not necessarily be exhibited in the natural setting, the hybrids would do well enough in the wild so that any genes they possess – including transgenes — would be passed on to wild radishes. Once in the wild populations, the new genes would be selected for if they provided advantages to wild radishes. This might happen, for example, if the new gene conferred resistance to insects. Even the transgenes provided no disadvantage, but were merely neutral; they might nevertheless be retained in the wild population.

However, in plants, desirable new phenotypes created by the introduction of foreign DNA are frequently unstable. This leads to a loss of the newly acquired traits and to a differential response to environmental conditions. In the case of petunias with an introduced maize AI gene, the plants produced different-colored flowers depending on the climatic conditions (light intensity and temperature) and on when the seeds were harvested. Greenhouse plants demonstrated less variability. In addition, plants grown from seed produced late in the growing season demonstrated a higher degree of instability.[31]

The risk of adverse environmental effects due to genetic material escaping from an engineered crop is thought to involve three major paths. Two of these, widespread pollen dispersal and the likelihood of cross-fertilization, are closely related. Escape via seeds or seed banks is a third possibility. Although no direct data on transgenic plants are available, available data using "natural" plants could provide a basis for examining these issues.

The effect of gene location and order has been studied by a number of workers in small-grain cereals[28] (reviewed by Whitkus et al.[73]). The effect of location and order has been noted since many related species have the same gene but exhibit different phenotypic effects. Use of the RFLP technique will aid in comparative analysis and permit elucidation of the mechanisms which control phenotypic expression.

3.2   Animals

The use of engineered organisms in the agricultural industry presents problems related to escape of higher organisms into the environment. Scientists are concerned with the potential for escape of microorganisms from laboratories and fields. However, higher animals are prone to actively escape. Usually higher transgenic animals are raised in much smaller numbers than microorganisms and multiply much slower. However, they can become established in the wild and interbreed with wild species.

Transgenic animals and microorganisms are used today for research, but also for production of pharmaceuticals (goats-lactoferrin)[40] and biological control agents.[23] Pigs and rabbits[10] are engineered to be organ donors for humans, and chickens have been used both as a model for vertebrate development and for applied purposes, like producing foreign proteins in hens' eggs.[47] Transgenes are expressed in over 25% of the offspring of pigs.[48]

Relative to animal biotechnology, Pursell and Rexroad[62] feel that "transgenes for productivity traits could substantially increase world food supply; however, these traits are controlled by numerous genes, only a few of which are presently known". Therefore, "Pharm" applications — agriculturally produced pharmaceutical products — will be on the market sooner than productivity-related agricultural applications of biotechnology because "insufficient information on gene regulation is available so that these integrated transgenes can be regulated", with the result that the overall health status of the animal will be adversely affected. For example, "considerable data now shows that the use of genomic DNA instead of cDNA is highly advantageous for obtaining high levels of expression" and a high incidence of gene transfer. The work of Brinster et al.[11] showed that the presence of introns increases transcriptional efficiency in transgenic mice.

Progress in the transgenic modification of swine and sheep has been reviewed recently by Pursell and Rexroad.[62] They described production of transgenic sheep expressing valuable pharmaceutical products in milk. The trait is passed on to the offspring. All animals appeared healthy. However, in mouse experiments, loss of ability to produce milk at an earlier than expected stage was associated with the presence of foreign genetic material. In contrast, production of swine with leaner meat or more rapid growth inevitably resulted in significant adverse effects on the animals.

Mogen et al.[51] point out that phosphorus as stored in seeds is a poor nutrient for monogastric animals. When phytase is present, phosphorus is released and is available as a nutrient. Their development of transgenic seeds containing the phytase gene from *Aspergillus niger* has led to an increase in the growth rate of chickens. This reduces the need to add phosphate to the feed, reducing costs and reducing excretion of phosphate to the environment.

Insertion of genetic material from multiple sources is being considered in the common sweet potato. Sweet potato is grown in more than 110 countries. China is the largest producer. Workers at Tuskeegee University[59] are developing procedures to increase disease and insect resistance by addition of genetic material from a variety of sources into the sweet potato. The use of viral coat protein and *Bacillus thuringiensis* genes is being explored. A synthetic gene has been produced which enhances the nutrient content of the potato and is being tested for expression.

3.3 Fish

The development of gene transfer as a means of improving cultured fish stocks is progressing rapidly, and use of transgenic fish in aquaculture seems possible within the next decade. It is likely that some transgenic individuals will escape into natural systems; the nature and extent of subsequent impacts upon native stocks and aquatic communities are presently unknown. Identification of likely mechanisms of ecological impacts and key gaps in knowledge needed to predict the extent of ecological risk associated with using transgenic fish in aquaculture is needed to develop appropriate risk assessment procedures.[35]

The development of transgenic fish introduces ecological and public policy uncertainties which may constrain the rate of progress towards practical utilization. Pathways giving rise to ecological impacts of transgenic fish in natural communities may prove complex and the magnitude of such impacts difficult to predict. Continuing uncertainty over policies regulating environmental release and commercial use has slowed the pace of development of genetically modified organisms. Knowledge of the benefits and risks of using transgenic fish, effective exercise of regulatory authority, and public confidence in each of these will prove important in determining whether and how soon transgenic fish reach practical utilization.[36,42,79]

Production of transgenic fish has been reviewed.[35,55] Hallerman and Kapuscinski[35] identified 13 species of fish to which foreign genetic material had been added; 4 years later, there is only one addition to the list (sea bream)[32] (Table 2). The characteristics altered remain basically the same:

TABLE 2

Species of Fish to which a Foreign
Genetic Material had been Added

Species	
Atlantic salmon	Mud carp
Channel catfish	Northern pike
Common carp	Rainbow trout
Goldfish	Silver crucian
Loach	Nile tilapia
Medaka	Zebra fish
Sea bream	

attempts to improve the rate of growth, resistance to stress, and addition of marker genes to ascertain if a transfer procedure was effective. Since research on gene transfer in fish was initiated in 1984,[37] transgenic fish have been produced by laboratories in at least 15 countries. A total of 96 genes have been cloned for transfer.[10]

Recent interest in pollution control has led to attempts to develop sentinel fish which could serve as indicators or early warning systems for pollutants. Winn and Beneden[76] have reported attempts in fish to include reporter genes that will be expressed in response to specific pollutants. In addition, interest in shellfish has resulted in attempts to develop transgenic oysters which will mature at a faster rate.[22]

In addition, recombinant DNA techniques have been applied to the production of vaccines[68] and diagnostic techniques.[4] Livestock vaccination is a normal part of cattle raising and is recognized as a release of the vaccinating agent.

Transfer of novel genes into fish, and the potential environmental effects and effects on consumers, as well as the use of recombinant vaccines, introduces a number of contentious issues into public policy debate among fisheries scientists and regulatory authorities. Anticipated ecological impacts of releasing such fish into natural environments have been described and reviewed.[37,79] The major determinant of ecological impacts of transgenic fish will be the phenotypic effect of the inserted genes. Hallerman and Kapuscinski[37] point out three conceptual classes of phenotypic changes that can be anticipated: changes in physiology, behavior, or tolerance of environmental factors. Zilinskas and Lundgrin[79] suggest a series of concerns:

1. disruption of local fauna
2. genetic degradation of overall gene pool by elimination of wild species
3. introduction of diseases new to the locale
4. significant degradation of the local environment
5. ability to eliminate the introduced organism

Hallerman and Kapuscinski suggested a position statement on transgenic fish for adoption by the American Fisheries Society which takes into consideration the science and safety issues and recognizes the political issues relating to legal jurisdiction.

Marine biotechnology itself offers little information that might answer or clarify questions about environmental safety or risk because its activities have so far been limited to laboratory research, with one exception: rearing transgenic carp in outdoor ponds. However, we may be able to derive data by examining aquatic biology, specifically to appraise the effects of past dispersals of exotic marine organisms into new environments. For example, the accidental releases of farmed salmon in the Pacific Northwest and in Europe may provide insights for risk assessment if the releases and effects can be documented and categorized. Other examples of successful dispersals (intentional or unintentional) of marine organisms and unsuccessful intentional attempts to introduce species will yield case studies which can be used to assess potential effects of introduced biotechnology products. This will enable an assessment of the impact of the safety issue on the advancement of marine biotechnology.

The continuity of oceans, the movement of water, and aquatic commercial activities favor dispersal of marine species, whether by accident or design. Generally, a species is well adapted to its native habitat, so its individual members usually will die when carried away from it. Sometimes a species will be accidentally transported to a foreign site and will find an ecological niche in which it can establish itself. In most cases, especially in terrestrial situations,[66,67] no effect will be seen; in a few cases a beneficial effect occurs. In a very few cases, adverse effects are noted. If these are of a great magnitude (e.g., *Sargassum muticum* along the English and French coasts or *Hydrilla* in the Chesapeake Bay area, *Penaeus vannamei* in the Pacific Rim area, and zebra mussel in the U.S.) they are well reported. In cases of lesser significance the reports are not widely circulated. Awareness of the potential value of examining the history of dispersal of exotic species has increased interest in historical and less "glamorous" data. In 1991 there was a major conference on the subject which resulted in a major publication in 1992.[14,18,79]

Besides accidental introductions and transferrals, humans have deliberately transported marine species from their home territories to new sites for commercial or environmental (e.g., mosquito control) purposes. For example, the *African tilapia* was introduced to Asia and Latin America, and the black tiger shrimp (*Penaeus monodonl*) and the white shrimp (*P. orientalis*) were introduced to many Asian and some Latin American countries. Deliberate introductions have been done for purposes other than aquaculture. *Gambusia affinis* and *Lebistes reticulatus* were introduced into parts of the world where malaria is endemic in an effort to reduce mosquito populations.[79]

There is a rapidly growing body of literature addressing the causes and effects of damaging dispersals,[2,17,18,75] focusing on marine plants,[56] mollusks,[19] shrimp,[18] and finfish.[21] Other workers have examined the potential for gene exchange[12] and the ubiquity of viruses in marine waters.[77]

Development of public policies regulating the use of genetically modified organisms is currently underway. In the U.S., environmental release policies for fish are under the jurisdiction of the Office of Agricultural Biotechnology of the U.S. Department of Agriculture,[1,36] which is finalizing its policy. The U.K. working group on releases to the environment has prepared a draft of its fisheries policy. Both of these groups stress the science issues involved in decision making based on assessing the risk involved. The U.S. Biotechnology Working Group of the Council on Economic Competitiveness of the White House Office of Science and Technology Policy recently[60] recognized the science issues, stressed the competitiveness issue, and raised the question of socioeconomic impact of biotechnology products. Resolution of these differences will have great bearing on the rate of progress and ecological safety of field tests with transgenic fish in the U.S. State governments are also undertaking development of biotechnology regulations, primarily in response to persistent gaps in environmental protection caused by unresolved policy conflicts at the federal levels.[37,79]

4   RISKS AND BENEFITS

Pest control using biological toxins is a significant part of transgenic research. In 1992, 7.6% of the trials involved genetic alteration of traits associated with pest control. *Bacillus thuringiensis* (BT) toxin genes have been inserted into other bacteria, modified and inserted into BT, and inserted into plants. The development of insect populations resistant to BT as a result of exposure to low levels of the toxin has been detected and researchers are developing new genes. The genes are combinations of insecticidal proteins toxic to lepidopterans but with different specificities. The hybrid toxin binds to different receptor sites, thus providing alternatives in resistance management studies.[9] Other means of inhibiting pests are being explored. Shade et al.[65] reported that transformed peas (i.e., added α-amylase inhibitor gene from the common bean) yielded seeds which could be stored which are resistant to infestation by the Bruchid beetle. Since the mode of action of this inhibitor is to cause modest mortality, prolong larval development time, and reduce fecundity, it may result in lowered development of resistance by insect populations.

Information presented at the Risk Assessment Symposium[64] indicated the importance of BT toxin interaction with clays. The toxin is known to be tightly bound to clays with little desorption. In addition, clay-bound toxin is more resistant to degradation in soil than free toxin. The exact mode of degradation is not known. The structure, antigenic properties, and insecticidal activity of the toxin are retained when it is bound to clay. When BT toxin was added directly to the soil or in transgenic cotton plants it was shown that the half-life in soil ranged from 24–41 days (when added directly) to 2 days (added as part of cotton plant).[41]

There is intense interest in the use of molecular biology techniques to produce plants which are resistant to diseases transmitted by viruses. Plant virus diseases cause major crop losses at high economic cost. This interest is evident in that 24% of U.S. field tests and 16% of field tests in Europe in 1993 were issued for plants in which virus resistance had been introduced.

The possibility of adverse ecological effects as a result of insertion of coat protein genes to induce natural plant protective mechanisms as protection against future infections has been suggested. Pleiotropic effects might occur as a result of the gene insertion. Relative to more direct effects of the inserted genetic material, the possibility of transcapsidation has been reported. Transcapsidation can occur when a second virus infects an already infected plant. If/when both viruses mature, both coat proteins are present, and the resulting virus particles may have an outer coat made up of a mixture of elements. This will confer on the resultant transcapsidated virus the ability to infect a different range of plants. Such mixed infections are common in nature.

Other phenomena of importance for risk assessment include the possibility of recombination through template switching or heterologous recombination, the potential synergism of mixed infections, and the role of helper-dependent (RNA) complexes in aphid transmission. All of these could occur naturally as a result of mixed infections.

Hassler[38] reported on a meeting held by the American Institute of Wine and Food to discuss biotechnology food safety issues. It was pointed out that disease resistance and insect resistance will raise yields and thus reduce acreage needed to produce foodstuffs in required quantities. The case in point was squash. Farmers often plant twice as much acreage as needed because of the well-known effects of viruses. The use of virus-resistant squash (being developed by Asgrow Seeds, most likely to be the second genetically engineered crop to reach the marketplace) will permit use of half of the land, pesticides, and herbicides with the same yield to the farmer. The reduction will result in less pesticide and herbicide utilization. The majority of field tests in 1992 (58%) involved herbicide resistance genes.

In order to understand and describe the fate of genetic material in a plant (or its product) which is commercially distributed, one must have specific information about the introduced material, the vector, and the delivery system. Specific data are needed to describe the fate of released material from the transgenic plant. Since an understanding of the biological fate of the genetic material and its produce will be important in predicting the behavior of the plant material vis-à-vis the environment, a complete description and understanding of the specifics of each released plant is necessary. In many cases, examination of the produce, insert, and vector descriptions should suffice. However, issues will differ on a case by case basis. For example, issues which may arise concerning the release of a plant with alterations in lipid content will not be relevant when considering plants with additional pesticidal activity.

Examples included two types of transformations: engineering of *Brassica napus* canola to produce and accumulate stearate[53] and *Bacillus thuringiensis* (BT) toxin in cotton plants.[41] The *Brassica napus* data indicated that weediness, persistence, invasiveness, selective advantage, pollen movement, outcrossing to wild relatives, and fate of the hybrids would not present significant environmental problems. The effect of altered stearate content on edibility to wildfowl was seen as a possible, but unlikely, problem.

One of the risk assessment issues most often raised is the use of antibiotic resistance genes as markers in organisms to be released. Bryant and Leather[16] have reviewed environmental safety issues and the need for alternate marker genes. Alternate means of selecting organisms expressing or containing the inserted DNA are being developed. Perl et al.[57] demonstrated the use of feedback inhibition in the aspartate family biosynthetic pathway as a means of selecting transformed plants. Plants containing either dihydropicolinate synthase or desensitized aspartate kinase genes derived from bacteria overproduce lysine and are therefore not affected by the exposure to toxic lysine analogues during growth.

5 RISK ANALYSIS FOR VACCINES

Rinderpest is the most important disease of livestock in developing countries of Africa and Asia. Unchecked, the virus can wipe out the cattle population in an entire country. In addition to the loss of cattle, quarantine efforts necessary to control the disease deprive poor countries of important sources of income.

Conventional rinderpest vaccines exist and have been highly effective in initially reducing incidence of the disease, but practical problems of follow-up and surveillance usually have led to the resurgence of the disease. Effective control of rinderpest would bring tremendous benefits to cattle-dependent populations of Africa. The U.S. Agency for International Development (AID) has issued its first approval for the testing of a recombinant animal vaccine in a developing country. The vaccine was engineered in the United States with AID funds — a vaccinia virus used to immunize human populations against smallpox. The recombinant vaccinia virus confers immunity to rinderpest.

According to documents provided by AID, the Agency has approved a small-scale, confined test that will be conducted over a 5-year period in the high-containment facility of the Kenya Agricultural Institute in Mugaga. The test will be done under the supervision of the Kenyan government and the Organization for African Unity.[3] But the recombinant vaccine is not without risk. The vaccinia virus on which it is based has an impressive safety record in healthy individuals, but it can cause disease and even death in immune-compromised individuals. The high prevalence of acquired immune deficiency syndrome (AIDS) in Central Africa makes this an important concern, especially for eventual use of the vaccine in mass immunization programs. Other safety concerns stem from the possibility of the recombination of the vaccinia virus with other naturally occurring pox viruses. Table 3 identifies vaccines tested and the risk assessment issues addressed.

One of the more successful and larger-scale field trials was performed in Belgium with a recombinant vaccinia-rabies vaccine.[13] The group has constructed recombinant vaccinia virus VVTGgRAB, expressing the surface glycoprotein (G) of rabies virus. The recombinant was a highly effective vaccine in experimental animals, in captive foxes, and in raccoons. The group reported the results of a large-scale campaign of fox vaccination in a 2200 km² region of southern Belgium, an area in which rabies was prevalent. After distribution, 81% of foxes inspected were positive for tetracycline, a biomarker included in the vaccine bait, and, other than one rabid fox detected close to the periphery of the treated area, no

TABLE 3

Examples of Risk Analyses for Experimental Veterinary Vaccines

Name of vaccine	Type of vaccine	Purpose
Pseudo-rabies (modified live virus)	Recombinant	Environmental release
Salmonella choleraesuis (avirulent live culture)	Conventional	Environmental release
Rabies (vaccinia vector)	Recombinant	Environmental release
Toxoplasmosis (modified live protozoa)	Conventional	Environmental release
Newcastle disease — fowl pox (fowl pox vector)	Recombinant	Environmental release
Rinderpest (vaccinia vector)	Recombinant	Contained study

* Copies of risk analyses are available for public inspection by contacting Dr. Jeanette Greenberg, Veterinary Biologies, BBEP, APHIS, USDA, Room 571, Federal Building, 6505 Belcrest Road, Hyattsville, MD 20782; phone (301)436-5390; fax (301)436-8669.

case of rabies, either in foxes or in domestic livestock, has been reported in the area.

A new risk analysis model[33] has been used to evaluate several field tests (Table 3) and the analyses are available to the public for review.* The Animal and Plant Health Inspection Service (APHIS) hopes this risk analysis model may be extended to analyzing other biotechnology drugs in the future.

6 THE DEVELOPING WORLD

6.1 Status of Regulations

Safe use of agriculture biotechnology products around the world requires region- or country-specific regulations. Many countries in the industrial world and the European Union have created at least some regulatory oversight of biotechnology. That is not the case, however, in most of the developing world. Lacking financial resources and scientific expertise, most developing countries have yet to establish agricultural biotechnology regulatory programs. Two recent publications on international agricultural biotechnology identify a number of countries in Africa, the Caribbean, and Latin America that lack enforceable government biosafety regulations (see list below).[61,72]

Examples of developing countries lacking
biotechnological regulations

Botswana	Mozambique
Columbia	Namibia
Costa Rica	Peru
Egypt	South Africa
Guatemala	Swaziland
Indonesia	Tanzania
Jamaica	Uganda
Kenya	Zambia
Malawi	

7 APPLIED RECOMBINANT RESEARCH IN DEVELOPING COUNTRIES

Nevertheless, in addition to working on traditional crops, developing nations are using biotechnology to improve native species, including sweet potatoes, papaya, banana, and palm — the plants on which many substance farmers depend. Richard Sawyer, President of the International Fund for Agricultural Research in Arlington, Virginia, said that in the West, researchers and farmers "have a cereal mentality, but there are alternatives crop, such as fruit and vegetables, that are of more interest to small farmers in developing nations".

Papaya, a staple of tropical diets, is infected almost worldwide by the ringspot virus, an RNA virus that is transmitted by aphids to papaya, squash, and related plants and greatly reduces their yields. About 7 years ago, Dennis Gonsalves of Cornell University's Agricultural Experimental Station in Geneva, New York began a collaboration with Richard Manshardt and Maureen Fitch of the University of Hawaii and Jerry Slightom of Upjohn to see whether papaya plants genetically engineered to express ringspot viral coat proteins could be resistant to the virus. Field trials with the genetically modified plants began in 1992. The results, Gonsalves said, were dramatic: they had 100% protection.

A problem remains, however. The protection conferred by the coat-protein gene appears to be specific to the viral strain from which the gene came. This may mean that transgenic papaya plants will have to be specifically tailored to resist strains indigenous to the areas where they will be grown.

Papaya isn't the only developing-world crop that's being modified to protect it from disease. International Services for the Acquisition of Biotechnology Applications, a not-for-profit international organization based at Cornell University that encourages the transfer of agricultural biotechnology to developing countries, is funding efforts by Marto Valdez and Gabriel Macayo of the University of Costa Rica to genetically engineer resistance to viruses into several varieties of the criollo melon, grown by small farmers in Costa Rica, Mexico, and Guatemala. Also, Magdy Madkour and his colleagues at the Agricultural and Environmental Research Institute are using transgenic techniques to improve disease resistance and nutritional qualities of the fava bean, a food eaten in many Mediterranean countries.[50]

But for many important crops, including cereal grains, beans, and peas, the danger from pests does not end with the harvest. Weevils and other insects may cause losses just as great — or greater — during storage. The problem is particularly acute in developing countries, where farmers can rarely afford protective chemical fumigants. Brazilian farmers, for instance, lose a staggering 20% to 40% of their beans to storage pests, leading to periodic food shortages.

Infestations of stored legume seeds by bruchid beetles, such as the cowpea weevil and the Azuki bean weevil, cause substantial economic and nutritional losses of these food crops, especially in developing countries. Seeds of the common bean are resistant to these bruchids largely because of the presence of α-amylase inhibitor (αAl-Pv), a seed protein that is toxic to the larvae. The αAl-Pv gene is therefore a candidate for a genetic engineering approach that would make other legumes (pea, chickpea, cowpea, Azuki bean) resistant to bruchid infestations. A team led by molecular biologist Maarten Chrispeels of the University of California, San Diego (UCSD), plant biologist T.J. Higgins of CSIRO's Division of Plant Industry in Canberra, Australia, and entomologist Larry Murdock of Purdue University in West Lafayette, Indiana tested this possibility by transforming peas (*Pisum sativum*) with the αAl-Pv gene driven by a seed-specific promoter. The levels of αAl protein in the pea seeds were as high as in bean seeds, and the peas were resistant to the cowpea and Azuki bean weevils.[65]

Field releases of GMOs in China were reviewed by Chen at the 2nd International Symposium on the Biosafety Results of Field Tests of Genetically Modified Plants and Microorganisms.[20]

Due to the rapidly increasing population and decreasing farming lands, China greatly needs biotechnology. Much progress has been achieved in biotechnology research with strong support from the Chinese government and some private foundations, such as the Rockefeller Foundation. Recombinant bacteria with higher efficiency of N_2 fixation have been released to about one million hectares of rice and soybean fields: higher yields of rice and soybean production have been reported, particularly in the fields lacking fertilizers. Transgenic plants have been developed: the coat protein genes of several Chinese strains of plant viruses such as tobacco mosaic virus, cucumber mosaic virus, potato virus X and Y, etc. have been cloned, and some coat protein genes have been introduced into tobacco, tomato, and some other economically important plants. Genes encoding the toxin of *Bacillus thuringiensis* have also been introduced into plants, resulting in insect-resistant transgenic plants. Transgenic tobacco and tomato plants resistant to virus infection have been released to the fields for testing the virus resistance performance and the stability of the foreign genes in transgenic plants in different environments. The size of the testing field reached over 200 hectares in 1991, and this makes China one of the largest countries in the world for field releases of transgenic plants.

8   CURRENT REGULATION OF INDUSTRIAL RECOMBINANT DNA

On July 18, 1991, the U.S. National Institutes of Health (NIH) added a section to its Guidelines entitled Good Large-Scale Practice (GLSP).[34] Similar to the Organization for Economic Cooperation and Development (OECD) version[24] from which it is derived, GLSP is a "level of physical containment that is recommended for large-scale research or production involving viable, nonpathogenic, and nontoxigenic recombinant strains derived from host organisms that have an extended history of safe large-scale use. Likewise, it is recommended for organisms such as those included in Appendix C that have built-in environmental limitations that permit optimum growth in the large-scale setting but limited survival without adverse consequences in the environment."[13] Highlights of this section of Appendix K include requirements for:

1. formulating and implementing institutional codes of practices to assure adequate control of health and safety matters

2. writing instructions and training personnel to assure that cultures are handled prudently and the workplace is kept clean and orderly

3. providing facilities, clothing, equipment, and protocols that are appropriate for the risk of exposure to viable organisms

4. assuring that environmental discharges are handled in accordance with applicable governmental environmental regulations

5. controlling aerosols in a manner such that employee exposure to viable organisms is at a level that does not adversely affect the health and safety of employees

6. including provisions for handling spills in the facility's emergency response plan

A comparison of GLSP with BL1-LS yields a number of differences that are summarized in Table 4.

GLSP is a less stringent level of containment for large-scale production of recombinant organisms which fall under the definition in Appendix K.

The implementation of GLSP, particularly those elements relaxing the requirements for filtration of exhaust, have the potential for saving the fermentation industry millions of dollars in capital investment. This comes without increasing the risk of injury to employees or the public or contamination of the environment.

This complements the blueprint for regulation of biotechnology in the U.S. (Coordinated Framework for Regulation of Biotechnology),[63,71] in which the jurisdiction of each federal agency is established.

Simultaneous activities in Europe included the adoption of three directives by the European Economic Community: "on the protection of workers from risk related to exposure to biological agents at work",[26] "on the contained use of genetically modified organisms",[24] and "on the deliberate release of genetically modified organisms".[25]

There has been some attempt at harmonization between the U.S. federal agencies and their European Community (EC) counterparts. For example, the Department of Transportation has accepted the United

Nations and International Air Transport Association requirements for transportation of hazardous materials and the Food and Drug Administration (FDA) is working with its counterpart in the EC.

As the U.S. strives to compete in the EC market, the issue of European standards presents a formidable challenge. The EC may require more stringent testing than the U.S. in the prized EC market. This becomes particularly frustrating when one considers that the U.S. can voice opinions in the standardization process but cannot vote. Ultimately, this process could result in European standards becoming global standards, which the U.S. would have to meet. The EC has already begun to establish standards on laboratory categorization; waste handling, inactivation and testing; codes of good practice for laboratory operations; guidelines for animal containment in experiments; definition of equipment needed in microbiological laboratories according to hazard; codes of practice for large-scale processing and production; standards required for quality control procedures; standards related to modified organisms for plant and soil application; and standards relating to microorganisms that are human, plant, and animal pathogens. The standards will define in concrete terms the technical specifications, codes, methods of analysis, and list of organisms needed to complement legislation. Unfortunately, they also will undoubtedly take some of the flexibility out of compliance with the directives. One of the benefits is that with the development of EC standards, the member countries cannot introduce their own standards in these areas.

Since the hazards of recombinant DNA (rDNA) technology were first discussed nearly two decades ago, the issue of regulation of this science has remained ever present. To date, the most successful approach has been the use of the NIH Guidelines — a document with no legal status which was initially conservative and cautious but which has become more lenient and less restrictive with time as the perceived hazards of rDNA are discovered to be unfounded. While many agree that regulation of biotechnology in some form is necessary, many feel that the lack of problems associated with rDNA technology thus far does not support the extensive amount of information required for testing and commercialization of GMOs in the EC. To prevent unnecessary crippling of this fledgling industry, regulation must be based on reality rather than on perception. Biosafety guidelines have been developed for the containment of pathogens and are effective regardless of whether or not the pathogen is a GMO. Once a realistic assessment is conducted for the relative risk of the agent, the activity, and the host, appropriate levels of physical and biological containment can be prescribed. Such facts must be incorporated into current and future regulations to assure that they reflect the actual risk rather than the public perception of risk.

TABLE 4

Comparison of GILsp and BL1/LS

Characteristic	GILsp	BL1/LS
1. Sealing of facilities		
handling of exhaust gases	Minimize leakage	Sealed containers
performance of adjusting valves	Minimize leakage	Equipment designed to prevent leakage
2. Conditions of work area		
biohazard sign	Not necessary	Necessary
air lock	Not necessary	Not necessary
decontamination and washing facilities	Optional	Necessary
shower facility	Not necessary	Not necessary
disposal facilities for waste water from decontamination, washing and shower facilities	Optional	Necessary
3. Ventilation	Optional	Necessary
maintenance of air pressure negative to atmosphere	Not necessary	Not necessary
application of HEPA filter to ventilation	Not necessary	Not necessary
design of work area to prevent contents from spreading outside the area in case spillage occurs	Not necessary	Necessary
design of work area to enable sterilization by fumigation	Not necessary	Necessary

9 INDUSTRIAL COUNTRIES

In the industrial world, herbicide-tolerant crops dominate the field trials of genetically engineered plants in the countries of the Organization for Economic Cooperation and Development (OECD) (see Table 5). A summary of a recent OECD study[30] shows that herbicide-tolerant crops were the subject of 57% of the approvals of field trials conducted in OECD countries since 1986. The number of approvals of herbicide-tolerant crops is more than four times greater than the number for the next most frequently tested trait — virus resistance. Next in frequency are potato (14%), tobacco (8%), tomato (8%), corn (8%), flax (6%), soybean (5%), and cotton (4%). Table 5 details the number and percentage of approvals for six traits by country. The total number of approvals is 846. The U.S. and Canada have approved the most field tests, together accounting for nearly three quarters of the OECD total. Oilseed rape is the most frequently tested crop, receiving 34% of the approvals.

TABLE 5

Number and Percentage of Approvals of Field Trials of
Engineered Plants for Traits and by Country

	Number of approvals*	Percent of total
Trait		
Herbicide tolerance	483	57
Virus resistance	113	13
Insect resistance	87	10
Quality traits	68	8
Male sterility	39	5
Disease resistance	35	4
Others	18	2
Country		
United States	316	37
Canada	302	36
France	77	9
Belgium	62	7
United Kingdom	45	5
Netherlands	22	3
Spain	6	<1
Sweden	6	<1
Denmark	3	<1
Germany	2	<1
Switzerland	2	<1
Australia	1	<1
Japan	1	<1
Norway	1	<1

* The meaning of an approval may vary from country to country. An approval may be for a test in a single site or multiple sites for one year or several years. The sum of approvals for all traits is greater than the number of approvals because some trials include crops with more than one genetically engineered trait. The number of field trials is greater than the number of approvals because one approval may cover several trials to be conducted at different sites and/or during different years.

REFERENCES

1. Agricultural Biotechnology Research Advisory Committee, Working Group on Aquatic Biotechnology and Environmental Safety. 1992. Minutes of Meeting on October 15, 1992. U.S. Department of Agriculture, Washington, D.C.
2. Allen, S.K., Jr. 1992. Issues and opportunities with inbred and polyploid species. In Introductions and Transfers of Marine Species: Achieving a Balance Between Economic Development and Resource Protection, DeVoe, M.R., Ed., South Carolina Sea Grant Symposium, Hilton Head, SC.

3. Anon. 1994. AID test of Riderpest vaccine in Kenya. *Gene Exchange*, 5 June.

4. Aoki, T. 1991. The application of biotechnology in fish pathology in Japan. In Proc. Int. Marine Biotechnology Conf., Baltimore, MD, p. 55.

5. Baker, H. 1972. Migration of weeds. In *Taxonomy, Phytogeography, and Evolution*, Valentine, D., Ed., Academic Press, London, 327.

6. Barrett, S. 1983. Crop mimicry in weeds. *Econ. Bot.* 37:255.

7. Bogosian, G. and Kane, J.K. 1991. Fate of recombinant *Escherichia coli* K-12 strains in the environment. *Adv. Appl. Microbiol.* 36:87.

8. Bogosian, G., Morris, P.J.L., Weber, D.B., and Kane, J.F. 1993. Potential for gene transfer from recombinant *Escherichia coli* K-12 used in bovine somatotropin production to indigenous bacteria in river water. *J. Ind. Microbiol.* 11:235.

9. Bosch, D., Schipper, D., van der Kleij, H., Maagd, R.A., and Stiekema, W.J. 1994. Recombinant *Bacillus thuringiensis* crystal proteins with new properties. *Bio/Technology* 12:915–919.

10. Brem, G. and Muller, M. 1994. Large transgenic animals. In *Animals with Novel Genes*, Maclean, N., Ed., Cambridge University Press, London, pp. 179–245.

11. Brinster, R.L., Allen, J.M., Behringer, R.R., Gelinas, R.E., and Palmiter, R.D. 1985. Introns increase transcriptional efficiency in transgenic mice. *Proc. Natl. Acad. Sci. U.S.A.* 82:4438–4442.

12. Britschgi, T.B. and Giovannoni, S.J. 1991. Phylogenic analysis of a natural bacterioplankton population by rRNA gene cloning and sequencing. *Appl. Environ. Microbiol.* 57:1707–1713.

13. Brochier, B., Kieny, M.P., Costy, F., Coppens, P., Bauduin, B., Lecocq, J.P., Languet, B., Chappuis, G., Desmettre, P., Afiademanyo, K., Libois, R., and Pastoret, P.P. 1991. Large-scale eradication of rabies using recombinant vaccinia-rabies vaccine. *Nature* 354:520.

14. Brock, J.A. 1992. Procedural requirements for marine species introductions into and out of Hawaii. In Introductions and Transfers of Marine Species: Achieving a Balance Between Economic Development and Resource Protection, DeVoe, M.R., Ed., South Carolina Sea Grant Consortium, Hilton Head Island, SC, pp. 51–54.

15. Brunken, J., de Wet, J., and Harlan, J. 1977. The morphology and domestication of pearl millet. *Econ. Bot.* 31:163.

16. Bryant, T. and Leather, S. 1991. Removal of selectable marker genes from transgenic plants: needless sophistication or social necessity. *Trends Biotechnol.* 10:274–275.

17. Carlton, J.T. 1989. Man's role in changing the face of the ocean: biological invasion and implications for conservation of near-shore environments. *Conserv. Biol.* 3(3):265–273.

18. Carlton, J.T. and Geller, J.B. 1993. Ecological roulette: the global transport of nonindigenous marine organisms. *Science* 261:78–82.

19. Cembella, A. and Shumway, S.E. 1994. Sequestering and biotransformation of paralytic shellfish toxins in scallops: food safety implications of harvesting wild stocks and aquaculture biotechnology. In Proceedings of the Symposium on Aquatic Biotechnology and Food Safety, OECD Secretariat, Ed., Organization for Economic Cooperation and Development, Paris, pp. 69–78.

20. Chen, Z.L. 1992. Field released of recombinant bacteria and transgenic plants in China. In Proceedings of the 2nd International Symposium on the Biosafety Results of Field Tests of Genetically Modified Plants and Microorganisms, Casper, R. and Landsmann, J., Eds., Biologische Bundesanstalt für Land und Forstwirtschaft, Brauschweig, Germany.

21. Chen, T.T., Lin, C.M., Gonzalez-Villasenor, L.I., Dunham, R., Powers, D.A., and Zhu, Z. 1992. Fish genetic engineering: a novel approach in aquaculture. In Dispersal of Living Organisms into Aquatic Ecosystems, Rosenfield, A. and Mann, R., Eds., Maryland Sea Grant College, College Park, MD, pp. 265–280.

22. Chen, T.T. 1994. Enhancing the growth rate of the eastern oyster. Aquaculture at the Center of Marine Biotechnology. UMBI PD9301. Baltimore, MD.

23. Cory, J.S., Hirst, M.L., Williams, T., Hails, R.S., Goulson, D., Green, B.M., Carty, T.M., Possee, R.D., Cayley, P.J., and Bishop, D.H.L. 1994. Field trial of a genetically improved baculovirus insecticide. *Nature* 370:138.

24. Council of the European Communities. 1990. Council Directive of 23 April 1990 on the contained use of genetically modified micro-organisms. 90/219/EEC. *Official J. Eur. Comm.*, No. L 117, 1.

25. Council of the European Communities. 1990. Council Directive of 23 April 1990 on the deliberate release into the environment of genetically modified organisms. 90/220/EEC, *Official J. Eur. Comm.*, No. L 117, 15.

26. Council of the European Communities. 1991. Draft Proposal for a Council Directive Amending Directive 90/679/EEC on the Protection of Workers from Risk Related to Exposure to Biological Agents at Work. Doc. No. 4903 EN.

27. Darmency, H. 1994. The impact of hybrids between genetically modified crop plants and their related species: introgression and weediness. *Mol. Ecol.* 3:37.

28. Devos K., Millan, T., and Gale, M.D. 1993. Comparative RFLP maps of homologous group 2 chromosomes of wheat, rye, and barley. *Theor. Appl. Genet.* 83:931–939.

29. Doebley, J. 1990. Molecular evidence for gene flow among *Zea* species. *BioScience* 40:443.

30. Evaluation of biosafety information gathered during field released of GMO's [DSTI/STPBS(92)6]. 1993. A Summary in Clearinghouse on Biotechnology, Friends of the Earth European Coordination, Mail-out No. 16, January, pp. 7–11.

31. Finnegan, J. and McElroy, D. 1994. Transgene inactivation. Plants fight back. *Bio/Technology* 12:883–890.

32. Funkenstein, B., Cavari, B., Moav, B., Harari, O., and Chen, T.T. 1991. Gene transfer of growth hormone in the Gilthead Sea bream and characterization of its pregrowth hormone. In Proc. Int. Marine Biotechnology Conf., Baltimore, MD, p.93.

33. Gay, C.G. 1994. A risk analysis model for experimental veterinary vaccines. A new way of analyzing risk is now available on disk. *Bio/Technology* 12:826.

34. Good large-scale practice (GLSP). 1991. In Appendix K of the Guidelines for Research Involving Recombinant DNA Molecules. U.S. National Institutes of Health, Bethesda, MD, July 18.

35. Hallerman, E.M. and Kapuscinski, A.R. 1990. Transgenic fish and public policy: regulatory concerns. *Fish* 15(1):12–20.

36. Hallerman, E.M. and Kapuscinski, A.R. 1991. Recent developments in public policies regulating the development of transgenic fishes. In Proc. Int. Marine Biotechnology Conf. (IMBC '91), Baltimore, MD, 13–16 Oct.

37. Hallerman, E.M. and Kapuscinski, A.R. 1992. Ecological and regulatory uncertainties associated with transgenic fish. In *Transgenic Fish*, Hew, C.L. and Fletcher, G.L., Eds., Singapore World-Scientific, Singapore, pp. 209–228.

38. Hassler, S. 1994. Not science, but necessary. *Bio/Technology* 12:7.

39. Heitkamp, M.A., Kane, J.F., Morris, P.J.L., Bianchini, M.D., Hale, M.D., and Bogosian, G. 1993. Fate in sewage of a recombinant *Escherichia coli* K-12 strain used in commercial production of bovine somatotropin. *J. Indust. Microbiol.* 11, 243.

40. Hyde, B. 1993. Down on the pharm. *ASM News* 59:115.

41. Jepson, P.C., Croft, B.C., and Pratt, G.E. 1994. Test systems to determine the ecological risks posed by deliberate release from *Bacillus thuringiensis* genes in crop plants. *Mol. Ecol.* 3(1):81–90.

42. Kapuscinski, A.R. and Hallerman, E.M. 1990. Transgenic fish and public policy: anticipating environmental impacts of transgenic fish. *Fish* 15(1):2–12.

43. Kareiva, P., Morris, W., and Jacobi, C.M. 1994. Studying and managing the risk of cross fertilization between transgenic crops and wild relatives. *Mol. Ecol.* 3(1):15–23.

44. Kidd, G. 1993. Commentary on chickenfeed. *Bio/Technology* 11:552.

45. Klinger, T. and Ellstrand, N. 1994. Engineered genes in wild populations: fitness of weed-crop hybrids of *Raphanus sativus*. *Ecol. Appl.* 4:117.

46. Linder, C.R. and Schmidt, J. 1994. Assessing the risk of transgene escape through time and crop wild hybrid persistence. *Mol. Ecol.* 3(1):23–71.

47. Love, J., Gribbin, C., Mather, C., and Sang, H. 1994. Transgenic birds by DNA micro-injection. *Bio/Technology* 12:60.

48. Maclean, N. 1994. Transgenic animals in perspective. In *Animals with Novel Genes*, Maclean, N., Ed., Cambridge University Press, London, 1994, pp. 1–21.

49. Maclean, N. and Rahman, A. 1994. Transgenic fish. In *Animals with Novel Genes*, Maclean, N., Ed., Cambridge University Press, London, 1994, pp. 63–106.

50. Moffat, A.S. 1994. Developing nations adapt biotech for own needs. *Science* 265:186.

51. Mogen, N.V., Verwoerd, T.C., van Paridon, P.A., van den Elzen, P.J.M., Geerse, K., van der Klis, J.D., Verseegh, H.A.J., van Ooyen, A.J.J., and Hoekema, A. 1993. Phytase containing transgenic seeds as a novel feed additive for improved phosphorous utilization. *Bio/Technology* 11:811–814.

52. Moore, G., Gale, M.D., and Flavell, R.B. 1993. Molecular analysis of small grain cereal genomes: current status and prospects. *Bio/Technology* 11:584–589.

53. Morra, M.J. 1994. Assessing the impact of transgenic plant products on soil organisms. *Mol. Ecol.* 3(1):53–57.

54. Muth, W.L., Counter, F.T., Richardson, K.K., and Fisher, L.F. 1993. *Escherichia coli* K-12 does not colonize the gastrointestinal tract of Fischer-344 rats. *J. Ind. Microbiol.* 11:253.

55. National Agricultural Library. 1993. Transgenic Fish Research: A Bibliography. USDA Number 117. National Agricultural Library, Beltsville, MD.

56. Neushul, M., Amsler, C.D., Reed, D.C., and Lewis, R.J. 1992. Introduction of marine plants for aqua culture purposes. In Dispersal of Living Organisms into Aquatic Ecosystems, Rosenfield, A. and Mann, R., Eds. Maryland Sea Grant College, College Park, MD, pp. 103–138.

57. Perl, A., Galili, S., Shaul, O., Bentzvi, I., Galili, G. 1993. Bacterial dihydropicolinate synthase and desensitized aspartate kinase: two novel markers for plant transforma-tion. *Bio/Technology* 11:715–720.

58. Pimentel, D., Hunter, M.S., Lagro, J.A., Efroymson, R.A., Landers, J.C., Mervis, F.T., McCarthy, C.A., and Boyd, A.E. 1989. Benefits and risks of genetic engineering in agriculture. *BioScience* 30:606.

59. Prakash, C.S. 1994. Biotechnological approaches to sweet potato improvement. *Biolink* 2(1):5–7.

60. President's Council on Competitiveness. 1991. Report on National Biotechnology Policy. Government Printing Office, Washington, D.C.

61. Proceedings of the USAID Latin America Caribbean Region Biosafety Workshop, Agricultural Biotechnology for Sustainable Productivity Project. Michigan State University, East Lansing, May 10–13, 1993.

62. Pursell, V.G. and Rexroad, C.E. 1993. Recent progress in the transgenic modification of swine and sheep. *Mol. Rep. Dev.* 36:251–254.

63. U.S. Department of Agriculture. 1991. Recombinant DNA molecules. *Fed. Regist.* 53:43411.

64. Seidler, R.J. and Levin, M.A. 1994. Potential ecological and nontarget effects of transgenic plant gene products on agriculture, silviculture and natural ecosystems. *Mol. Ecol.* 3(1):1–5.

65. Shade, R.E., Schroeder, H.E., Pueyo, J.J., Tabe, L.M., Murdock, L.L., Higgins, T.J.V., and Chrispeels, M.J. 1994. Transgenic pea seeds expressing the alpha amylase inhibi-tor of the common bean are resistant to bruchid beetles. *Bio/Technology* 12:793–796.

66. Sharples, F.E. 1991. Ecological aspects of hazard identification. In *Risk Assessment in Genetic Engineering*, Levin, M.A. and Strauss, H., Eds., McGraw-Hill, New York.

67. Simberloff, D. 1985. Predicting ecological effects of novel entities. In *Engineered Organisms in the Environment*, Halvorsen, H.O., Pramer, D., and Rogul, M., Eds., American Society for Microbiology, Washington, D.C.

68. Thiry, M., Dheur, I., Xhonneux, F., Margineau, I., Dommnes, L., Vanderheijen, J., Rossius, M., Kinekelin, P., and Renard, A. 1991. Vaccination against fish rhabdovirus. In Proc. Int. Marine Biotechnology Conf., Baltimore, MD, p. 56.

69. Tiedje, J.M., Colwell, R.K., Grossman, Y.L., Hodson, R.E., Lenski, R.E., Mack, R.N., and Regal, P.J. 1989. The planned introduction of genetically engineered organisms: ecological considerations and recommendations. *Ecology* 70:298.

70. Till-Bottraud, I., Reboud, X., Brabant, P., Lefranc, M., Rherissi, B., Vedel, F., and Darmency, H. 1992. Outcrossing and hybridization in wild and cultivated foxtail millets: consequences for the release of transgenic crops. *Theor. Appl. Genet.* 83:940.

71. United States National Institutes for Health. 1991. Guidelines for Research Involving Recombinant DNA Molecules. National Institutes for Health, Bethesda, MD.

72. Visser, B. 1993. Biosafety Conference: a step towards regional co-operation in South Africa. *Biotechnology and Development Monitor*, #17, December.

73. Whitkus, R., Doebly, J., and Lee, M. 1992. Comparative genome mapping of sorghum and maize. *Genetics* 132:1119–1130.

74. Wilkes, H. 1977. Hybridization of maize and teosinte in Mexico and Guatemala and the improvement of maize. *Econ. Bot.* 31:254.

75. Williamson, P. and Gribbin, J. 1991. How plankton change the climate. *New Sci.* 129:48–52.

76. Winn, R.N. and Beneden, R.J. 1991. Development of transgenic fish for study of aquatic contaminants. In Proc. Int. Marine Biotechnology Conf., Baltimore, MD, p. 78.

77. Wommack, K.E., Hill, R.T., Kessel, M., Russek-Cohen, E., and Colwell, R.R. 1992. Distribution of viruses in the Chesapeake Bay. Appl. Environ. Microbiol. 58:2965–2970.

78. Yancey, R.J., Jr., Kotarski, S.F., Thurn, K.K., Lepley, R.A., and Mott, J.E. 1993. Absence of persistence and transfer of genetic material by recombinant *Escherichia coli* in conventional, antibiotic-treated mice. *J. Ind. Microbiol.* 11:259.

79. Zilinskas, R.A. and Lundgrin, C.G. 1993. Marine Biotechnology and Developing Countries. World Bank Discussion Paper 210. World Bank, Washington, D.C.

Chapter 3

Biological Safety Considerations for Environmental Release of Transgenic Organisms and Plants

Daniel F. Liberman, Linda Wolfe, Richard Fink, and Elizabeth Gilman

TABLE OF CONTENTS

0-8493-4465-4/96/$0.00+$.50

1 INTRODUCTION

Genetically modified organisms for use in agriculture promise sub-stantial benefits for the production of food for the world's people. Scien-tists are developing potatos that contain more solids, strawberries that can retain their texture when frozen, vine-ripened tomatoes that last for months, and vegetables that can be produced in the absence of chemical pesticides and fertilizers. Yet for various reasons these products have engendered worldwide discussion concerning potential risks and adverse socioeco-nomic impacts that may be associated with the introduction of transgenic products into the marketplace. While there may be issues that should be addressed by the industry and government regulators during product development and testing, the present consensus is that any risks associ-ated with these products will not be different from risks associated with conventional products of a similar nature and, therefore, should be man-ageable by familiar methods.[2] This same conclusion was reached in con-sidering the occupational (human health) hazards associated with genetic engineering.[16]

2 HISTORICAL BACKGROUND

The application of these technologies to agriculture represents both a breakthrough and a continuation of historic practice. Modern agriculture is derived from centuries of manipulation of plants and animals. Nearly all the crop species now grown commercially have been artificially bred.[12] These hybrid varieties are produced by the mixing and matching of the best traits produced through random mutation to create useful combina-tions. Creation of these new varieties of living organisms has depended on a form of classic genetic engineering practiced for centuries, in contrast to the more precise methods of cellular and molecular manipulation now available. This classic approach operates at the level of the organism. The end result, improved organisms for man's benefit, is the same!

The advent of the newer approaches has given scientists the ability to isolate genes from any source and to transfer them to other agriculturally useful organisms. Scientists are no longer limited by species boundaries. They have an ever-increasing pool of useful genes to explore. The potential is there for dramatic new solutions to existing agricultural problems, as well as the creation of new crops tailored for use all over the world, especially for underdeveloped countries where growth of traditional crops has proved unsuccessful.[14,20]

These studies are very complex and a great deal of care must be exercised to ensure that they are successful. Scientists do not simply create random "new" forms of life and rush into field trials to determine how they will behave in the environment. They proceed in an orderly sequence which will define those parameters significant to evaluating their potential utility. This involves a continuum of testing from research laboratory to greenhouse or glasshouse prior to introduction into the environment. Controlled experiments are conducted in properly designed facilities prior to release.[9,11] Each stage of experimentation may be revisited several times to obtain reproducible data. It may be necessary to construct and reconstruct organisms with better field performance. It is clear that transgenic plants and microorganisms are extremely powerful and useful tools, and we need to learn how to harness the energy and excitement associated with their use.

3 REGULATORY CONSIDERATIONS

Codes of good practice or guidelines which address occupational health and safety in the laboratory, greenhouse, or glasshouse have been established to ensure safe conduct of research in the initial research and development effort (see NIH-52, *Fed. Regist.* 154: 29800–29813). Unfortunately, similar codes of practice and safe operating principles for environmental safety have not been compiled for the basic and applied small-scale open research.[11]

Several federal government agencies have statutory authority for regulating field trials and biotech products. The U.S. Department of Agriculture (USDA) and the Environmental Protection Agency (USEPA) have authority over agricultural biotechnology activities (field trials). The USDA's regulatory powers are exercised by the Animal and Plant Health Inspection Service (APHIS) (Plant Pest Act; 7 U.S.C. 150 aa–jj). USEPA regulatory authority derives from two specific laws: the Federal Insecticide,

Fungicide, and Rodenticide Act (FIFRA; 7 U.S.C. 136–136y) and the Toxic Substances Control Act (TSCA; 5 U.S.C. 2601–2929). FIFRA is the law under which the EPA regulates all pesticides. The agency has used this law in the past to register a number of naturally occurring microbes and viruses for use as biological pest control agents. The USEPA is using existing regulations to review and register genetically engineered microbial pesticides intended for commercial use. However, this agency instituted a policy in 1984 under FIFRA (49CFR 50886) to extend its authority to certain small-scale field tests of genetically engineered biopesticides which would otherwise have been exempt from agency review. Under this authority, the USEPA has reviewed dozens of small-scale field test proposals.

Commercial activity in agricultural biotechnology is regulated by the Food and Drug Administration (FDA) (also see Coordinated Framework of Biotechnology Science Coordinating Committee), when biotechnology products are intended for use in human food or animal feed.

Under the Food, Drug, and Cosmetic Act (21 U.S.C. 301–392), the FDA is responsible for both preventing "adulterated" foods from being sold and approving the use of specific food additives. The agency also classifies certain foods and food additives as "generally recognized as safe" (GRAS) and has regulations establishing criteria for all these classifications.

The last agency that has regulatory authority in this arena is the Occupational Safety and Health Administration (OSHA). OSHA has made it very clear that it would treat biotechnology manufacturing activities similar to the way conventional processes are regulated (50 *Fed. Regist.* 14468–14469). Although very little has been said or written about regulation of farm workers' occupational exposure to biotechnology products, it can be expected that government regulatory policies will be similar to existing worker safety regulations, since the new products will be handled and used in much the same way as existing products.[2,11]

The authorities of the various federal agencies were carefully described in the Federal Register (*Fed. Regist.* 31118–31121). This publication established the principle for federal oversight whereby representatives from each agency would assess safety and environmental impact prior to obtaining field trial approval from either the USDA or EPA.

4   ENVIRONMENTAL CONSIDERATIONS

With regard to naturally occurring or classically mutated microbes (generated without the benefit of recombinant DNA [rDNA] technology), the greatest number of potential environmental effects may be associated with the use of transgenic microorganisms in the environment due to persistance and spread.[13] The safe use of these in agriculture and other environmental applications for decades, however, provides some assurance that similar uses of rDNA engineered organisms will likewise be safe. The most commonly used microbes are the nitrogen-fixing soil bacteria called rhizobia, which have been used as seed additives for legume crops since the 1890s. Other bacteria have been safely used as biological pesticides, primarily *Bacillus thuringiensis*, which produces a protein that is toxic only to insects. The history of use of these organisms indicates that they can be manufactured and handled safely and that they can be used in the environment with no adverse effects.[5,8]

These assertions can be supported by experience, both anecdotal and experimental, particularly for an organism like *Rhizobium*, which is one of the best studied commercially useful microorganisms.[3] Rhizobia are naturally associated with plant root nodules. They form symbiotic relationships with plants like alfalfa, clover, and legumes which develop these nodules, a good example of biological containment.[3] This may not necessarily apply to soil organisms with different ecological behavior (e.g., remediation bacteria which metabolize chemical soil contaminants).

Evidence of its safety derives from the decades in which it has been used by farmers and from extensive literature available on the environmental behavior of rhizobia — for example, its persistence in soil.[2,15] Rhizobia are not native to U.S. soils but became established upon the introduction of their associated legumes. Significant populations are only found in fields where the host legume has recently been grown,[11] and even then populations exhibit a cyclical pattern, with the greatest numbers in the spring and very low levels during winter.[20,26] The mobility of these bacteria in soil is known to be quite low, although limited dispersal is possible through wind, rain, etc.[3,8] Rhizobia survive poorly in groundwater or sewage and are known to fare poorly under desiccation or exposure to sunlight.[15,21,26] Data of this sort, coupled with the known lack of pathogenicity or toxicity, support the contention that the environmental use of rhizobia is safe.

Less information is available for other organisms; however, it is widely believed that most small-scale environmental uses of genetically engineered microorganisms will present little risk. Recent review articles by Kelman et. al.[14] and Tiedje et. al.[26] provide excellent summaries of relevant literature in this field and offer the basis for future risk assessments.

5   PRINCIPLES FOR RISK ASSESSMENT

In order to define a set of general experimental principles under which small-scale field research of low or negligible risk can be conducted with a specific transgenic material, it is essential that the following working assumptions be evaluated (see Reference 22):

1. Certain general scientific principles related to the organism, the research site, and experimental conditions have relative importance in determining whether an experiment is of low or negligible risk.

2. Conclusions regarding the risk of an experiment can be reached by evaluating the relevant factors and their interaction under the conditions of the experiment, including (this is to include evaluation), when available, existing data from greenhouse and laboratory studies.

3. Interactions of these factors are easier to address in small-scale field experiments than in large-scale experiments because of their limited scope, thus permitting closer monitoring, generally easier assessment and analysis, and the possibility of more effective containment measures in the event of unforeseen and potentially damaging occurrences.

4. There are a number of factors, such as the characteristics of the organism(s) and genetic material, used in the research site and the surrounding environment.

5. Experimental conditions must be evaluated in determining the safety of any specific experiment.

Under a broad range of conditions, the use of certain organisms may pose little or negligible risk. Other organisms with known adverse effects may be acceptable for field experiments if the design of the trial (i.e., control of experimental methods, confinement of the organism or its genetic material to research site) reduces the likelihood that these adverse effects will occur. The term "site" is intended to include the research plot proper and an appropriate part of the surrounding environment. The risk associated with the activity can be reduced by choosing a site comparable to one in which a similar study has documented whether dissemination and establishment beyond the site occurred. In addition, the investigator should be able to choose a suitable safe research site by identifying important ecological and/or environmental considerations relative to safety in the specific geographical location (e.g., high water table, heavy field run-off, etc.), climatic conditions, size of physical area, and an appropriate geographical location in relation to proximity to specific biota that could be affected.[9,14,22,25]

How do you approach assessing risk rationally? One approach could be to compare the risk posed by the introduction of transgenic organisms to previous introductions of similar organisms in similar target/test environments. In this way if an introduction can be shown to be similar to a previous introduction, one which was shown to be safe, then the level of risk should be similar as well. For certain introductions there may be insufficient knowledge that existing practices or regulations adequately address the possible risk posed by the introduction.

In a rational world, the need for extensive oversight for well-characterized systems, regardless of the method of organism modification, would be unnecessary. However, we do not live in a rational world. We live in a world where public opinion, public reaction, and public acceptance are the keys. Considering this, it becomes important to factor public perceptions into the decision to release an rDNA-modified organism because that is where the issue(s) are decided.

The importance of this preception is supported by the results of a recent study in which 35 researchers and institutional regulatory affairs professionals were interviewed concerning their experiences with field trials.[13] This survey indicated the following:

1. Federal regulation of biotechnology in general, and field tests in particular, is necessary given current levels of public concern about safety of biotechnology research.
2. Biosafety protocols currently required for field tests are overly cautious. However, they are necessary to reassure the public.
3. Permitting agencies have shown flexibility in adjusting safety requirements commensurate with the field testing experience of various organisms.
4. Those who are involved with field tests are acutely sensitive to public concerns and go to considerable lengths to provide information about the tests to the press, to officials at all levels, and to the public at large.
5. There is an increase in the regulation of biotechnology at the state level, often leading to costly duplication of effort and delays.

Representatives from the large companies indicated that if this trend continues it could force even those companies to cut back on or abandon field testing. Several of those interviewed during the survey pointed out that while the coordinated framework can work reasonably well for field tests, food safety issues will become paramount as transgenic food products reach the commercialization stage.[13]

Those interviewed indicated that their pretest expectations for the behavior of the transgenic organisms and the safety of the field test were met with no surprises. The biomonitoring procedures utilized during the test also substantiated pretest predictions concerning the safety of the transgenic organisms. In some instances, meeting the biomonitoring requirements of the field test permit heavily taxed personnel and material resources and cut into efficacy analyses. One test failed because the plants selected were not cold resistant; in another instance, a worker mistakenly decapitated the plants, and in a third, the late discovery of a native plant variety necessitated an adjustment in the field test protocol.[13]

Many of those interviewed stated that while current field testing with small, tightly contained, isolated plots may be adequate for "Proof of

Concept" trials, the only valid determination of commercialized performance and safety of transgenic organisms will come through large-scale trials. One very interesting question that was asked in the survey was: Did your experiences with the project cause you to conclude that you might have done some aspects of the test differently? The overwhelming response to this question was that the initial test design had been adequate, but for subsequent tests, adjustments were made in plot size, planting time, genetic construct, amount of observation, etc. that would improve the efficacy of the test. The most common reaction to the question was that the biosafety elements of the test, particularly first-time tests, were overdone. For one first-year test with a transgenic tomato, the biosafety requirement called for strict fencing, intruder detectors, and a guard. The fruit had to be harvested green, cooked at 90°C, and incinerated. In the second application, the requirement was proposed and accepted by the agency to simply harvest all the fruit and bury it. For the third test, this was adjusted to harvesting the red fruit, leaving the unripened ones in the field for incorporation into the soil, and monitoring for reemerging plants.

With some reservations, most of those interviewed concluded that their biosafety experiences could be extended to other closely related organisms and to other genetic inserts. They felt that extension to transgenic plants would be mostly straightforward and should lead to streamlined approvals for previously tested systems. They felt that extension to microorganisms was somewhat problematic because microorganisms present greater uncertainty if less is known about their ecology.[15]

6   GENERAL ISSUES CONCERNING MICROORGANISMS

Even when environmental behavior for a microorganism that will be modified through genetic engineering is known, a case-by-case risk assessment of each proposed biotechnology activity may be required. For microorganisms, the following general issues are likely to be considered. This list is not meant to be exhaustive; rather it is presented to illustrate the extensive evaluation that may be appropriate.[2,11]

6.1   Does the Organism Have any Detrimental Effects on Plants, Animals, or Humans?

The field trial must consider the potential for the organism to exhibit toxic effects on humans or animals, pathogenic effects on other organisms, or other ecological effects on target or nontarget species. While some of this information may be available from the literature, investigators may need to conduct specific studies which address whether there are any risks involved from exposure of farm workers, manufacturing or seed processing employees, or others who may come into contact with the product.

Many of the microorganisms that are tested in field plots are designed to have effects on other target organisms. Research using unmodified microorganisms has been conducted with little adverse effect on the environment even though microorganisms are known to have effects on other organisms in the test environment. For decades, plant pathologists have used microorganisms that cause plant disease in the field to evaluate plants for disease resistance. Plant pathogens have been tested in the field to gain fundamental knowledge about the biology and the pathogenicity of these microorganisms. Microorganisms used as biological control agents are specifically selected or modified to affect a target pest organism.[1] Some microorganisms, such a *Bacillus thuringiensis*, are used routinely in the environment as biological control agents for some lepidopteran insects. The issues that are routinely examined in these tests are helpful in assessing criteria for testing genetically modified microorganisms.

When the impact of a microorganism is studied, it is important to evaluate the effectiveness on the target organism and any side effects on nontarget organisms. Microorganisms that are engineered to act as biological control agents may have genes inserted which encode toxins; they may have an extended host range or may have increased virulence for a particular target organism. The effect of any new trait, especially changes in the host range of the microorganism, should be evaluated before field testing. Potential nontarget organisms should be identified by experimenting with representative species under contained conditions. It is unlikely that the relative abundance of a species in a community or ecosystem will be significantly altered as a consequence of small-scale field research if the microorganism can be effectively limited to the plot and its immediate surroundings. Yet it is important that field research be conducted so as to limit exposure to sensitive nontarget species.[2]

These concepts can be applied to specific examples. New strains of *B. thuringiensis* should be experimented with on a plot on which no threatened or endangered species of lepidopteran insect will be exposed. It is essential that great care be taken in testing beneficial insects for sensitivity to the test microorganism and in limiting the exposure of significant populations of sensitive beneficial insects.[2]

6.2   Does the Organism Persist in the Environment?

It is essential to define the extent to which the transgenic organism may persist in the soil, water, etc. Although persistence itself may not be an undesirable trait, it is useful to confirm the prediction that introduced organisms will, or will not, compete poorly in new environments and will, or will not, multiply out of control.

The probability that a released organism will become established is influenced by a number of population parameters (e.g., number of organisms, distribution, size, habitat, susceptibility, basic biology, etc.).

The survival of test organisms is dependent on the numbers applied and the reproductive success of the organisms in the test site. In order to increase its numbers the experimental microorganism must be able to either (1) compete effectively against other organisms in the research site or (2) find within the test site a niche without competitors or one that contains less effective competitors. The probability that an introduced microorganism will be an effective competitor is dependent on the source of the test organism and whether or not the engineered strain has any advantage in the environment in which the test will occur. In some instances, the microorganism will be reintroduced into an environment from which it or its parental microorganism was isolated. In such situations, neither the introduced gene nor the introduced microorganism will be new or unique in that environment. What might differ is the frequency at which the gene/microorganism combination occurs in that site (see, e.g., the studies on *Pseudomonas syringae*[16,17]).

The modified organisms would then be in competition with the indigenous microorganisms. A knowledge of both the function of the added gene and the behavior of the parental organisms can be useful in predicting the response of the modified organism to factors such as competition for nutrients, predation, environmental stress, selection, etc.

The bottom line is that only through actual field trials can behavior be assessed. Therefore, the competitive ability of the experimental microorganism may have to be tested in the laboratory, greenhouse, glasshouse, or microcosms. Such studies are very important in predicting outcomes of small-scale field trials.

7   SUSCEPTIBILITY

Microorganisms that require a host for their survival are referred to as host-obligate or host-dependent microorganisms. Most available information concerning the factors which affect the competitive ability in host-dependent microorganisms comes from studies of microorganisms as biological control agents and from plant pathology breeding studies. In the microorganism/plant interaction, host-obligate microorganisms may be on the surface of the plant, inside plant tissues, or both.

Organisms which grow inside plant tissue (endophytes) have few competitors when compared to surface growers or free-living soil microorganisms.[1] While some plant viruses are known to survive in water, soil, and crop debris, viruses, viroids, and some prokaryotes (e.g., rickettsia-like bacteria, mycoplasmas, and spiroplasmas) which multiply entirely within their host generally do not survive when exposed to the outside environment. The environment in which they must grow and reproduce is determined to a great extent by the host. Although they may have fewer microbial competitors, these agents must deal with host defense mechanisms.[23]

Surface-growing obligate microorganisms (epiphytes) may be categorized on the basis of the kind of nutritional relationship they maintain with the host.[16] In their residency phase on leaves or roots, certain microorganisms obtain nutrients (as leaf or root exudates) from the plant but cause no harm to it. However, given the right conditions, they can kill and destroy host tissues through the action of toxins and enzymes. They then multiply in the dead tissue. In a second type of nutritional relationship, the organism obtains nutrients from a plant by killing the host tissue in advance of colonization.[23]

Many of the factors affecting competition among free-living soil microorganisms can be seen in host-dependent microorganisms. These include competition for space, competition for nutrients, selective predation, environmental stress, etc. In addition to dealing with these factors, the ability of these microorganisms to find suitable hosts in order to reproduce is a factor which can be used in designing an experimental protocol to test these organisms safely.

During the design of small-scale field trials the issue of density and distribution of potential habitats should be considered. Test site locations are reviewed on the basis of the distribution and size of likely potential habitats in the experimental region. The key is to select sites that provide very limited, if any, opportunities for disseminated organisms to find an appropriate habitat. If the density of potential habitats is low and the habitats are separated by relatively large distances, the probability of successfully establishing themselves is greatly reduced. Strategies based on habitat density are frequently used in agriculture to limit pathogen dissemination and crop destruction. For example, fields can be planted with several different varieties of the same crop species, each species possessing a different gene for resistance to the pathogen. Since the pathogen does not find a sufficient density of susceptible plants, it does not disseminate in an epidemic fashion. While a microorganism may proliferate within the experimental plot when such strategies are employed, it does not proliferate outside the plot because it does not find suitable hosts.[6,10,11,26]

7.1   Can the Organism Spread Beyond the Test Plot?

While dispersal itself may not be a negative occurrence, it is important to determine if an organism can spread beyond the test site. Although this often must be measured during the initial small-scale field trial, there may be evidence from studies with the parental organism that may allow predictions to be made.

Microorganisms are transported by a variety of routes, such as wind, water, mechanical means, and biological vectors.[4,5,8]

The effectiveness of dispersal generally depends on several factors. These include ability to adhere to soil or other particles, ability to infect vectors, ability to adhere to potential means of mechanical transport (e.g., animals, humans, and their tools), and ability to survive transport. These factors are dependent on the biological characteristics of the experimental organism. Therefore, biological characteristics of the test microorganism must be considered in evaluating the safety of field research.

While some microorganisms are dispersed by a variety of means, others may be restricted to one or a few modes of movement. In general, the more highly evolved a microorganism is to move by one route, the lower are its chances of moving by other routes. An understanding of potential routes of movement, as well as knowledge of methods of limiting its dispersal along these routes can be used to design safe field research and underline the need for monitoring.[6,7,10]

7.1.1   Wind Dispersal

The effectiveness of aerial dispersal (aerosols) is influenced by the following factors: mechanisms of entering the atmosphere (take-off), particle shape, ability to survive environmental stress (e.g., desiccation, UV light), and ability to adhere to soil and other particles.[4,5] Rafts of soil or dust particles are raised by wind when the ground is heated by solar radiation. The microorganisms attached to these soil particles are transported as the soil is blown by the wind. Some microorganisms adhere to insects or mites which can then be dispersed by wind currents. Some fungi explosively expel their conidia.[4] It is important to use the positioning of a field research plot to address and limit potential transport though the aerial route. For example, consideration can be given to natural features such as trees, hills, windbreaks, or fences which can be used to influence wind currents when experimental sites are identified. When the test microorganism possesses a high potential for dispersal by the aerial route, the positioning of the small-scale research plot on an off-shore island may be necessary to provide acceptable security.

7.1.2   Water Dispersal

In water, dispersal is influenced primarily by the transport properties of the suspending medium. Thus, the hydrology of soil water and groundwater flow and the proximity of open bodies of water such as lakes, rivers, streams, and water supplies for irrigation are among the primary physical determinants of waterborne dispersal from a terrestrial experimental plot.[14,19] Rain or irrigation water can also serve as a means of transport. Bacteria, viruses, spores, and mycelial fragments of fungi can be dispersed by rain or irrigation water that washes the surfaces of plants or moves over or through soil.[18,19] Rain splashes can throw microbe-laden droplets from plant surfaces into the air.[8]

The research plot can be designed to limit dispersal through these potential routes. For example, border strips around the research site can be used to isolate plants within the research plot. If necessary, the research plot can be situated so as to limit access of the test microorganism to groundwater or open bodies of water under both average and exceptional climatic conditions. On occasion it may be necessary to control water flow through the use of drainage, collection, and physical barriers.

7.1.3 Mechanical Dispersal

7.1.3.1 Human Activities

Organisms can be dispersed over short and long distances through the successive handling of plants, through the use of contaminated tools and other equipment, and through the transport of contaminated soil, plants, seeds, and nursery stock. Likewise, any activity that generates aerosols represents a potential route of dispersal for microorganisms contained in the aerosol droplet. Mechanical disturbances such as tilling introduce microorganisms into the air (along with the soil). Aerosolized soil may then settle downwind of the test plot. Mode of application of organisms varies. For small-scale field research, care should be taken to limit dispersal of microorganisms by human activities. Access to the test plot should be restricted to those individuals trained in procedures appropriate for limiting dispersal. Transport of microorganisms on contaminated materials should be restricted by use of appropriate decontamination procedures.

7.1.3.2 Animal Activities

In nature, animals can serve as vectors for microorganisms. Bacteria may be transported by browsing and burrowing mammals, soil arthropods, earthworms, and soil adhering to the feet. Insects can transport microorganisms. Their bodies can become coated with bacteria or fungal spores as they move among plants. Insects may carry microorganisms on the surfaces of their bodies and deposit them on plant surfaces or in the wounds that insects make on the plants during feeding. It may be necessary to limit the access of animals to small-scale test areas. Screening or fencing of the experimental site may be necessary to protect the integrity of the study. Maintenance of such physical barriers is essential; therefore, some form of continued monitoring may be necessary to ensure their effectiveness.

7.1.4 Biological Vector Dispersal

As noted before, microorganisms can be transmitted by insects during feeding and movement of the insect from plant to plant. The insect acts as a vector that carries the microorganism from one place to another where it can become established.

The relationship between the vector and the microorganism can be either persistent or nonpersistent. The persistent relationship occurs when the insect is able to transmit the microorganism over an extended period of time and the microorganism can multiply in the insect. Nonpersistence refers to a relationship in which the vector acquires the microorganism after a short feeding period on the plant, transmits the agent to another plant immediately after feeding, and then rapidly (within minutes) loses the microorganism.

Insects can transport microorganisms over both short and long distances. Other arthropods, such as mites, which cannot fly can be transported by wind. In this way dispersal of microorganisms over extremely long distances (hundreds of kilometers) has been shown.[22,24]

Care should be exercised in the design of small-scale field research trials. If it is known that the test microorganism can be transmitted by insects, then the trial should be located at an altitude where the vectors are not present. The test could also be performed at a time of year when the insect population is low. Alternatively, the use of insect repellents or denying vectors access to plants by netting can also be included in the experimental design. These arrangements are rarely totally effective in eliminating vector activity and are highly dependent on the climatic situation of the particular season concerned.

7.2 Is the Introduced Material Able to be Transferred (Gene Transfer) to Other Organisms?

Theoretically, genetic material or its encoded products that are safe in one host organism may function differently in another host organism. Therefore, if the introduced genes can be transferred to other species, there may be additional questions to raise in the risk assessment. Some genetic vectors, such as plasmids used in rDNA studies, are capable of transfer to different species, but the recent trend in both rDNA and agricultural biotechnology is to use nontransmissible vectors or to integrate new genes into the bacterial chromosome; both of these strategies dramatically reduce the risk of unwanted genetic transfer.[1] Clearly, the extent of genetic transfer should be assessed before field tests begin.

The gene transfer capability of the engineered microorganism or the stability of the genetic construct may affect interaction between microorganisms.

The factors to be considered in analyzing the effects of gene transfer on the safety of a genetically modified microorganism include the following:

1. What is the probability of horizontal transfer of the genetic material?
2. If the gene is transferred, will the new genetic information be maintained and expressed?
3. What is the known function of the new genetic material?
4. If the modified microorganism moves beyond the point of introduction, how will it affect, as a result of the transformation, the surrounding populations or communities of plants, animals, and indigenous microbes?

Gene transfer refers to the dissemination of genetic material through natural genetic mechanisms. The mechanisms by which plasmids and/or chromosomal genes are transferred include conjugation, transformation, transduction, and cell fusion. Although these mechanisms have been studied in the laboratory, few exchanges of genetic material in nature or in simulated natural settings have been documented. Little is known about the frequency of genetic exchange in nature. We expect genetic transfer frequencies to be lower in nature than laboratory frequencies.

Factors that affect transfer include bacterial densities (enhances mating frequency), the presence of free DNA (promotes transformation), the presence/absence of clay materials or minerals that may promote growth and plasmid transfer but not transduction, and host range. Additional factors such as the spatial, temporal, and physiological separation of bacteria; immobilization through adhesion to soil particles, organic materials, and other living organisms; genetic barriers such as restriction systems and plasmid incompatibility; and a variety of environmental conditions are also important. (See Reference 22 for an extended discussion.)

While estimates can be made of the transfer frequencies likely to be observed in specific environments based on laboratory studies, the probability that a genetic transfer is likely to occur and the significance of such transfers is hard to predict.

While questions such as these can be addressed in laboratory studies before small-scale field tests are done, the most reliable information comes from the small-scale tests themselves. Laboratory and greenhouse studies are not always reliable indicators of actual field behavior (crucial information can be site, temperature, and soil type specific for microbes released). Environmental monitoring must accompany small-scale field tests if the fate of the introduced microorganism is to be determined. These results can then be used to assess the potential risks when the product is used on a larger commercial scale.

This graduated testing approach should ensure that engineered microorganisms will be approved for commercial use only when the potential risks have been well studied. While there have been many field trials performed over the past 6 or 7 years, the results are not widely known. So far, genetically engineered microorganisms have shown limited dispersal from the field site and persistence in the environment.[9,13]

The risks associated with small-scale field research can be determined by analyzing the characteristics of the organism and the research site and by designing appropriate scientific and environmentally acceptable experimental conditions. Tests with microorganisms usually involve large populations, some portion of which may persist. When considering the safety of a field trial, microorganisms must be thought of in statistical terms that consider the probability of an event occurring in a given population/environment.

8 SCIENTIFIC CONSIDERATIONS FOR SMALL-SCALE FIELD RESEARCH WITH TRANSGENIC PLANTS

In much the same way that the FDA has defined Good Manufacturing Practice (GMP) for manufacturing and Good Laboratory Practice (GLP) for laboratory activities, there are basic practices which can facilitate the design and conduct of field trials.[22] The purpose of these "Good Field Practices" (GFP) would be to ensure that the transfer of genetic material of interest is controlled and that the dissemination of microorganisms or plant elements containing that genetic material is controlled.

These practices designed to prevent unintended, uncontrolled adverse effects on GFPs can be applied in a number of ways. For example, the experiment allows for control of transfer of genetic material and dissemination beyond the research site.

- In this case the biology of the organism could minimize the probability of horizontal gene transfer, or measures could be taken to prevent or minimize this gene transfer.
- The organism could have a limited ability to compete; thus, measures can be taken to minimize movement/dispersal of the microorganism from the test site.
- Measures can be taken to prevent or mitigate establishment beyond the test site if necessary.

If prior experience indicates that there is little likelihood of harm to (or significant impact on) areas beyond the research site:

- Results from contained studies or previous field trials demonstrate that there should be no adverse environmental effects beyond the research site, even if the microorganism should disseminate from the site.

- The experimental design is such that effects on other organisms (e.g., plant or animal health, microbial communities, ecosystem processes, other biological systems) can be readily detected and can be controlled or mitigated should they occur.

The ability of a microorganism or plant to disseminate into the environment and to transfer genetic material to other organisms and the availability of suitable, reachable habitats/niches in the vicinity of the research site will be important factors in evaluating safety.

Selective plant breeding has been practiced in some form for thousands of years. It was after the rediscovery of Gregor Mendel's work in 1900, however, that the systematic breeding now practiced by plant breeders became widely used. Observations made by scientists, based on a knowledge of plant genetics, plant morphology, plant reproductive biology, etc., have resulted in the current practices that breeders use to ensure the genetic integrity of their experimental material. This experience and that gained from the controlled field tests of genetically modified plants help to identify plant characteristics and experimental conditions that allow the safe conduct of small-scale field research.

The size of field experimental plots will more than likely be determined by the characteristics of the experimental plants; i.e., orchard crops will require larger experimental plots, while grain crops can be adequately evaluated using smaller experimental plots. Small-scale field research with genetically modified plants is conceptually analogous to the small-scale field research already conducted by plant breeders in evaluating potentially useful new varieties. The genetic modifications achieved through conventional plant-breeding techniques have produced single or multiple gene mutations and changes in chromosome number through chemical treatment or ionizing radiation, crosses between crop species, and interspecific crosses, including crosses between cultivated species and crosses between cultivated species and related noncultivated species.[19] When conducting conventional breeding research, attention is often given to preventing possible genetic influx from any sexually compatible plants into the research plot. It has not been demonstrated to date that natural transfer of genetic material from plants to organisms other than plants occurs.

Conventional small-scale field research evaluates the characteristics of a new plant variety and its interaction with the environment. Field experiments of new plant varieties produced by conventional plant-breeding methods have shown that most new plants resulting from breeding experiments are of no practical use to the breeder and are eliminated, with no further effect on either the environment or on subsequent plant breeding.[10]

There have been some instances where the intentional or accidental introduction of a foreign plant species into a new environment has had an adverse environmental impact. Examples (see Reference 24 for a discussion of these) include johnsongrass, introduced into South Carolina as a forage plant in the 1830s; water hyacinth, introduced into Florida as an aquatic ornament; and the Asian weed kudzu, introduced as a stabilizer of soil embankments and as crop forage on unproductive land. Many other important weeds (Canada thistle, yellow star thistle, field bindweed) now present in the United States are the result of the accidental introduction of foreign plant species. In Europe there have been similar problems as a result of intentional or accidental introduction of foreign plant species such as sunflower, common ragweed, and giant hogweed, which can cause severe dermatitis in man. These examples involve the release of naturally occurring nonindigenous plants rather than genetically modified plants. Therefore, the field testing of genetically modified plants conducted using GFP should not be considered analogous to uncontrolled introduction of foreign plants into entirely new environments.

9   REPRODUCTIVE ISOLATION OF GENETICALLY MODIFIED PLANTS

Employing practices that ensure reproductive isolation of the modified plants is an excellent method for preventing dissemination of genetic material from the test plant into other members of the same or related species.

In considering natural mechanisms for reproductive or genetic isolation one can identify characteristics that can be controlled by manipulating the experimental plants prior to mating.[1] Plants manipulated in this way can be made incapable of producing and/or disseminating any genetic material (via pollen, seeds, etc.) that would allow new genes to become permanently incorporated in the gene pool of the species.

Reproductive isolation is currently used by plant breeders and by seed producers to produce genetically pure seed. In these practices, the emphasis is on preventing the contamination of the test or breeding plants with extraneous genetic material (in most cases via pollen) in order to maintain the genetic purity of the experimental or breeding plant population.[11] The practices used to protect the genetic purity of a breeding line are similar to those used in field research (prevention of dispersal of the genetic material of experimental plants from the test plot), namely those based on the principle of reproductive isolation. These practices result in the spatial, mechanical, temporal, and genetic isolation that evolutionary biologists use to define reproductively isolated plant populations. In most cases the objectives of GFP would be achieved if field research with modified plants was designed to reproductively isolate them from the pool of sexually compatible plants outside the experimental site. In this way, small-scale field research could be conducted with reasonable assurance that it would not have significant adverse effects on the environment.

The list which follows illustrates the types of practices that are appropriate for reproductive isolation. The end result of using such practices should be that genetically modified plants are reproductively isolated.[11]

Examples of current experimental practices used to maintain reproductive isolation in plants include the following:

- Spatial separation is the most common method used to isolate sexually compatible plant populations. Most requirements for growing pedigreed or certified seed include some specification as to the distance the field must be from any field containing plants of the same species. The specific distance required will depend on the biology of the species in question. Self-pollinated species with fragile pollen will require relatively short distances, while some open-pollinated species with hardy pollen will experience some degree of contamination when separated from compatible plants by as much as several miles.

- In some instances, removal of the male or female reproductive structure(s) may allow plants to be safely grown in close proximity to compatible plants. An example of the use of this method is the mechanical removal of tassels in seed corn production. By removing the tassel (containing the pollen-producing male flowers) it is possible to eliminate entirely the source of genetic material from the male that can be transferred via pollen.

- A variation of the technique discussed above involves the incorporation into the plants in question of a cytoplasmic male sterility trait. When this trait is present, almost no viable pollen is produced, and the plant will virtually remain reproductively and biologically isolated.

- It may be possible to grow the plants in question in such a way that flowering will occur either earlier or later than it would be expected to occur in plants of nearby compatible crops and/or wild plant species. This use of temporal reproductive isolation can potentially be as effective as spatial separation in limiting the movement of genetic material.

- Pollen dissemination may also be prevented by physical means such as covering of flowers (bagging).

- When the objectives of a field test do not require seed production, as when forage qualities of alfalfa are being evaluated, it may be possible to harvest plants prior to flowering. In this case, reproductive isolation could be achieved in some crops that might otherwise be difficult to isolate.

Although reproductive isolation is likely to be the main safety concern for most small-scale field tests, there may be cases in which additional measures might be considered. For example, if the plants to be field tested have been modified to contain or express toxins or to contain biological vectors capable of transferring genetic material, then physical isolation may also be required.

10   EXPERIMENTS WITH PLANTS

The level of risk of small-scale field research with plants can be estimated by analyzing the characteristics of the organism and the research site and by designing appropriate, scientifically and environmentally acceptable experimental conditions. The prudent choice of research sites and experimental conditions coupled to field practices discussed previously is important in field research safety.

The plants most likely to be tested are domesticated crop species. In many cases there is extensive experience with their reproductive isolation and with the prevention of spread of plants outside the test area. Most domesticated crop plants cannot persist or thrive in noncultivated environments.

Characteristics of plants to be considered in experimental design include the biology of the reproductive elements of the plant, such as its flowers, pollination requirements and seed characteristics, and its history of controllable reproduction with lack of dissemination and establishment in an environment comparable to the research site; the mode of action, persistence, and degradation of any newly acquired toxic compound; the nature of biological vectors used in transferring DNA to plants; and any information on the interactions with other species and / or biological systems.

It should be clear that the field practices should facilitate the design and conduct of experiments. It should be possible to design the trial to reproductively isolate modified plants from sexually compatible plants outside the experimental site, to prevent genes or modified organisms from being released into the environment beyond the research site, and to use plants that, even without reproductive isolation, will not cause unintended, uncontrolled adverse effects.

11   CONCLUSION

The basic safety precaution is to confine experimental plants and microbes to the test plot area. Microorganisms can be contained by minimizing aerosols during application, using buffer crops to hinder soil mobility, preventing windborne dispersal by standard soil erosion control measures, and limiting animal and human access to the plot through appropriate security precautions. Initially tests genetically engineered plants should be tested in areas that are isolated from commercially important growing regions used in areas surrounded by buffer crops. (Glass, personal communication.) Pollen dissemination can be prevented by bagging flowers or through early harvest. These procedures are clearly not feasible for widespread larger-scale use. Bagging and early harvest will no longer be necessary for those products for which controlled small-scale field testing shows an absence of negative environmental effects.

The same is true for worker-protective clothing or equipment. Most genetically engineered agricultural products are not considered to be health risks. The need, therefore, for elaborate protective measures for the workers conducting field tests does not exist. Field tests of genetically engineered rhizobia and all plant field tests conducted to date have needed no such precautions.[13,18] There may be some instances in which initial testing of a biological pesticide may suggest cautionary use of protective clothing or where federal or state law may require such measures. Initially, state regulations required the now-infamous "moon suits" to be worn for the first field test of the "ice minus" frost protection microorganisms. Subsequent field tests were conducted with virtually no protective clothing, and only a simple hospital-style face mask was used for respiratory protection.[11,13]

It seems clear that government agencies will continue to follow the stepwise approach of reviewing each increase in scale before allowing commercial use.[21] It is anticipated that results of small-scale testing will be considered when assessing the risks of larger-scale uses. Products that show potentially adverse effects in small-scale testing will bear a greater regulatory burden before larger tests are allowed, if they are allowed at all. Finally, approvals for commercial use will be treated using existing regulations (e.g., for pesticides, foods, etc.) while also taking into account results from earlier field work.

The use of biotechnology has become subject to a number of excess regulatory requirements. What is overlooked is that the products of genetic manipulation in agriculture, new crop plants and plant-associated microorganisms, greatly resemble existing agricultural products produced by traditional means. These new products will be manufactured and sold in a manner identical to traditional products, using established distribution channels. Farmers and other workers will process and handle these products just as they handle existing products. In the final analysis, the risks posed by the new products of biotechnology will be no different than those of their traditional counterparts, and these risks will be regulated and managed in much the same way as those of existing agricultural products.

Scientifically acceptable and environmentally sound field research requires careful experimental design — e.g., formulation of a hypothesis and statement of objectives; development of specific methodologies for introduction of organisms, monitoring, and mitigation; a precise description of the design of experiments, including planting density and treatment patterns; and description of specific data to be collected and of methods for analysis to test for statistical significance.

Additional precautions may be required when considering a particular organism or trait or for particular environments (such as aquatic environments).

The following precautions should be considered in designing the trial:

1. Keep the numbers of the modified organism to the lowest level appropriate for the experiment.

2. Exercise measures to limit dispersal and establishment beyond the test site and supplement these measures when appropriate.

3. Continuously monitor the organism within the research site, both during the experiment and after termination. Be prepared to apply control or mitigation measures if appropriate and to avoid unintended adverse environmental effects during, at the termination of, or following the experiment.

4. Test for both the presence of established organisms and transferred genetic information outside of the primary research site.

5. Apply control or mitigation measures if appropriate and necessary to avoid adverse environmental effects outside of the primary research site.

6. Develop procedures for termination of the experiment and waste disposal.

7. Provide appropriate training on all necessary safeguards for all personnel involved in research.

8. Maintain records regarding the conduct of the trials and associated safety precautions.

If researchers design and conduct field experiments carefully, and if protocols for conducting small-scale field are carefully prepared and followed, then we will all enjoy the benefits of research with transgenic plants and microorganisms.

REFERENCES

1. Agrios, G. N.; *Plant Pathology*, 3rd Edition; Academic Press, New York; 1988.
2. Alexander, M.; Ecological consequences; *Issues Sci. Technol.* pp. 57-68; Spring, 1985.
3. Ames, P., Bergman, K.; Competitive advantage provided by bacterial mobility in the formation of nodules by *Rhizobium meliloti*; *J. Bacteriol.* 148:728–729; 1981.
4. Aylor, D. E., Waggoner, P. G.; Aerial dissemination of fungi spores; *Ann. NY Acad. Sci.* 353:116–122; 1980.

5. Banttari, E. E., Benette, J. R.; Aerosol spread of plant viruses — potential role in disease outbreaks; *Ann. N.Y. Acad. Sci.* 353:167–173; 1980.

6. Baum, R.; Field tests of recombinant organisms shown safe; *Chem. Eng. News* 67:30–34; 1989.

7. Betz, F., Levin, M., Rogul, M.; Safety aspects of genetically engineered microbial pesticides; Paper presented to the American Chemical Society, Washington, D.C.; September 1, 1983.

8. Bovallius, A., Roffey, R., Henningson, E.; Long-range transmission of bacteria; *Ann N.Y. Acad. Sci.* 353:186–209; 1980.

9. Center for Science Information; Regulatory Considerations, Genetically Engineered Plants; Summary of a Workshop held at Boyce Thompson Institute for Plant Research at Cornell University, Ithaca, NY; October 1987.

10. Gasser, C. S., Fraley, R. T.; Genetically engineering plants for crop improvement; *Science* 244:1293–1299; 1989.

11. Glass, D.; Agricultural biotechnology, occupational health and regulatory issues; *Occup. Med. State of the Art Rev.* 6:301–309; 1991.

12. Hardy, R. W. F., Glass, D. J.; Our investment — what is at stake?; *Issues Sci. Technol.* pp. 69–82; Spring 1985.

13. Hoban, T. J., Kendall, P. A.; Consumer Attitudes about the Use of Biotechnology in Agriculture and Food Production; Report to the Extension Service; U.S. Department of Agriculture, Washington, D.C.; 1992.

14. Kelman, A., Anderson, W., Galkow, S., et al.; *Introduction of Recombinant DNA Engineered Organisms into the Environment, Key Issues;* National Academy Press, Washington, D.C.; 1987.

15. Liang, L. N., Sinclair, J. L., Mallory, L. M., Alexander, M.; Fate in model ecosystems of microbial species of potential use in genetic engineering; *Appl. Environ. Microbiol.* 44:708–714; 1982.

16. Lindow, S. E., Knudsen, G. R., Seidler, R. J., Walter, M. V., et al.; Aerial dispersal and epiphytic survival of *Pseudomonas syrigae* during a pretest for the release of genetically engineered strains into the environment; *Appl. Environ. Microbiol.* 54:1557–1563;1988.

17. MacKenzie, D. R., Blowers, J. H.; The Federal Regulation of Biotechnology — A Report on a National Survey; Cooperative State Research Service; U.S. Department of Agriculture, Washington, D.C.; 1990.

18. Madsen, E. L., Alexander, M.; Transport of *Rhizobium* and *Pseudomonas* through soil; *Soil Sci. Soc. Am. J.* 46:557–560; 1982.

19. Mahler, R. L., Wollum, A. G.; Seasonal variation of *Rhizobium meliloti* in alfalfa hay and cultivated field in North Carolina; *Agron. J.* 74:428–431; 1982.

20. Mallory, L. M., Sinclair, L. L., Liang, L. N., et al.; Survival of microbial species used in recombinant DNA research and technology; in Abstracts of the Annual Meeting of the American Society for Microbiology; Washington, D.C.; Abstract #Q59, p. 219; 1982.

21. Office of Science Technology Policy; *Fed. Regist.* July 31, 31119; 1990.

22. Organization for Economic Co-operation and Development (OECD); Safety Considerations for Biotechnology; OECD Publications, Paris; 1992.

23. Reyes, V. G., Schmidt, E. L.; Population densities of *Rhizobium japonicum* strain 123 estimated directly in soil and rhizospheres; *Appl. Environ. Microbiol.* 37:854–858; 1979.

24. Sharples, F. S.; Spread of organisms with novel genotypes: thoughts from an ecological perspective; *Recomb. DNA Tech. Bull.* 6:43–56; 1983.

25. Sinclair, J. L., Alexander, M.; Role of resistance to starvation in bacterial survival in sewage and lake water; *Appl. Environ. Microbiol.* 48:410–415; 1984.

26. Tiedje, J. M., Colwell, R. K., Grossman, Y. L., et al.; The planned introduction of genetically engineered organisms: ecological consideration and recommendations; *Ecology* 70:297–315; 1989.

Chapter 4

BIOTECHNOLOGY: ITS APPLICATION TO WASTE MANAGEMENT

Sue Markland Day and Robert S. Burlage

TABLE OF CONTENTS

0-8493-4465-4/96/$0.00+$.50
© 1996 by CRC Press, Inc.

1 INTRODUCTION: GLOBAL HAZARDOUS WASTE MANAGEMENT PROBLEMS

In 1990/1991, the Roper Organization polled the American people on environmental issues.[1] The results of this study, *Environmental Protection in the 1990's: What the Public Wants*, found that, when compared to 1987, 22% more individuals believed that trying to improve the quality of our environment should be a major national effort. In fact, more individuals voted for the environment than for improving the quality of public school education. In this report, actively used and abandoned hazardous waste sites ranked as the most serious environmental problems.

As the size of the waste management and restoration problem is more fully documented, more public and political support is garnered for new cleanup technologies, and the U.S. waste management and restoration problems are large. For example, in the 1990 U.S. Environmental Protection Agency (USEPA) paper *The Nation's Hazardous Waste Management Program at a Crossroads: The RCRA Implementation Study*, 211,000 facilities were identified as generators of hazardous waste; each of these facilities is a potential site for contaminated soil and groundwater cleanups. The USEPA's 1988 *Report to Congress: Solid Waste Disposal in the United States* found that a quarter of the municipal landfills which had groundwater monitoring in place were leaking. At that time, 6,000 landfills had been counted nationwide. The USEPA's 1990 Toxic Release Inventory reported the release of 4.8 billion pounds of toxic chemicals annually, with 260 million pounds of hazardous waste produced each year. Moreover, as many as 10,000 to 5,000 oil spills occur each year and approximately 15% of the nation's underground storage tanks are leaking. A University of Tennessee study published in December 1991 estimated expenditures topping $750 billion over the next 30 years for U.S. remediation activities, and this figure did not include the cost of recurrent waste management activities.

The United States is not alone. Every year, the European Community[2] (EC) produces more than 2 billion tons of wastes, of which more than 21 million tons is dangerous.[3] For example, it has been estimated that as much as 50,000 acres of land in England is contaminated and that about 2% is so damaged that treatment is necessary before redevelopment.[4] In 1980, a survey of contaminated land in the Netherlands was undertaken in which more than 4,000 potentially contaminated sites were identified, with remedial actions recommended immediately for 350 sites. Sweden has also completed a national inventory of old dumps and landfills registering 3,800 sites. It is anticipated that 500 of these sites are candidates for remediation.

As in the United States, waste management and remediation is a big business in Europe. Two million employees and sales somewhere in the range of ECU 100 to 200 billion made up the EC waste treatment sector.[5] The Eastern European environmental market (equipment and services) has been projected by Frost and Sullivan, Inc. to approach $2.55 billion (U.S.) by the year 1995.[6]

Environmental protection programs and markets are also growing in the Pacific Rim countries. In 1990, ChinaTrade reported that Taiwan plans to increase its investments in pollution control from 3.41% gross national product (GNP) in 1988 to 8.73% GNP by the year 2000. Solid waste and wastewater treatment facilities are budgeted for $460 million and $1.1 billion, respectively. As one can see from the U.S., EC, Eastern Europe, and Taiwan figures, hazardous waste treatment, disposal, and contamination have global importance. To restore our polluted lands and waters and to integrate treatment at the point of waste generation are important international goals emphasized at the 1992 Earth Summit held in Brazil.

2   THE TASK AT HAND

"Waste management" can refer to the disposition of any material that is discarded by society, although for the purposes of this chapter "waste" refers to materials that are resistant to degradation. This is an important distinction, since most of the world's waste is organic material that is easily degraded by microorganisms and which then is recycled into the ecosystem.

Solid waste can be composted (an increasingly attractive option for many communities), while liquid waste (municipal sewage) is processed at a waste treatment plant or in a septic tank. Degradation-resistant waste, on the other hand, presents a challenge. Many of these chemicals which have been altered by man are not found in nature, and biochemical pathways to degrade them are uncommon. Some chemicals, such as heavy metals or radioactive elements, are toxic to living organisms. Other chemicals present a problem because of the large quantities that are made (and discarded). Small quantities of these chemicals might be degraded by indigenous organisms, but large quantities overwhelm the natural catabolic properties of a soil or groundwater community. The degradation of hazardous and recalcitrant chemical waste — and the uses of genetically engineered bacteria to accomplish this end — will be the focus of this chapter.

3   ONE SOLUTION: BIOREMEDIATION

The use of bioremediation to degrade hazardous waste has received a great deal of attention in recent years. One of the greatest attractions of bioremediation is that the waste material can be broken down to simple, harmless compounds or "mineralized". Sometimes the waste cannot be degraded further, as with heavy metals, but can be detoxified or immobilized to prevent further harm to the community. The organisms used are usually microorganisms (bacteria and fungi), although the use of algae, plants, and other organisms has been suggested. This chapter will concentrate on bacteria- and fungi-mediated applications, since most of the research and the relevant field examples involve these simple forms of life.

There is a vast amount of environmental biotechnology research underway, and a growing number of bioremediation field applications have been reported. The field of bioremediation is not sharply defined in the literature, but instead encompasses a great many treatment regimes. This variability makes discussions of bioremediation difficult until the scope of the problem is clearly defined. For instance, the contaminant of interest may be a single chemical species (e.g., toluene), a discrete group of closely related compounds (e.g., xylenes, PCBs), or a complex mixture of compounds (e.g., fuel oil). There may be heavy metals and other toxic chemicals in the contaminated matrix. On the biological scale, workers have described pure cultures of bacteria that degrade pollutants, but they have also described consortia that are completely undefined. Commercial applications of biology-based waste remediation may be performed *in situ* with indigenous organisms or may rely on the addition of bacteria or fungi, either recombinants or nonrecombinants, to a site. Finally, the physical scale may vary from bench-top reactors to pilot-scale facilities to engineering projects on a vast scale, such as *in situ* treatment of an aquifer or an oil spill in seawater. See Insert 1 for more detail on different physical designs for bioremediation technologies. Among these three variables (contaminant, organism, physical scale) a huge variety of projects may be described. See Insert 2 for advantages and Insert 3 for disadvantages of bioremediation.

INSERT I

Bioremediation Technologies Extracted from USEPA's VISITT Guidance[7]

Bioremediation — *In Situ* Groundwater. The defining characteristic of this technology is an injection system to circulate microorganisms, nutrients, and oxygen through contaminated aquifers. In most instances, groundwater is pumped, treated to some extent, and then reinjected with additives that enhance biodegradation. Biodegradation relies on contact between contaminants in the groundwater and microorganisms.

Bioremediation — *In Situ* Soil. The target media for this technology are subsurface soils and the vadose zone above the water table. In this technology, various microbes, nutrients, and an oxygen source are injected through injection wells into the soil. In general, subsurface soil moisture is required, and soils must be relatively permeable.

Bioremediation — Slurry Phase. This technology mixes excavated soil, sludge, or sediment with water to form a slurry that is mechanically agitated in an environment (usually a tank or reactor vessel, although *in situ* lagoon applications are possible) with appropriate ambient conditions of nutrients, oxygen, pH, and temperature. Upon completion of the process, the slurry is dewatered and the treated material discarded.

Bioremediation — Solid Phase. In this system, excavated soils are placed in a tank or building or on a lined treatment bed. Using convenient equipment, nutrients and other additives are tilled into the soil to facilitate microbial growth. The tillage equipment may provide aeration for the soil as well. Water is provided via a sprayer or sprinkler system. Composting and land farming are included in this category.

Note: Bioremediation is a technology which uses microorganisms to degrade organic contaminants or transform inorganic compounds. The microorganisms break down the organic contaminants by using them as a food source.

INSERT 2

ADVANTAGES OF BIOREMEDIATION

- It is ecologically sound, a natural process.
- The target chemicals are destroyed or detoxified, not merely transferred from one environmental medium to another.
- Biology-based waste treatment is reported to be less costly than other technologies, such as incineration.
- Bioremediation can often be accomplished where the problem is located, eliminating the need to transport large quantities of contaminated wastes off site.

INSERT 3

DISADVANTAGES OF BIOREMEDIATION[8]

- Research is needed to develop and engineer bioremediation technologies that are appropriate for sites with complex mixtures of contaminants.
- Cleanup using bioremediation often takes longer than other remedial actions, such as excavation or incineration.
- In some cases, depending on the contaminant, toxic by-products may be produced.

4 APPLICABILITY

It is convenient to separate bioremediation applications into two groups: (1) treatments for a site that is already contaminated and (2) treatments for a recurrent waste stream. The former group is represented by the Superfund sites, a list of sites that had been used in the United States as dumping grounds for hazardous waste over many years, and by those United Kingdom sites eligible for cleanup under the Derelict Land Grant program. The recurrent waste group reflects the ongoing problem that industries face when their manufacturing process generates a hazardous waste as a by-product. Both types of sites can benefit from bioremediation, although the engineering problems and the scale of the treatment process are usually quite different. For certain contaminated sites, bioremediation may be the only reasonable alternative. These sites would include contaminated subsurface soil and groundwater (aquifers). Such contamination cannot be physically moved to a processing site, and the land area or volume is too vast for any conventional treatment regime. Instead of bringing the contamination to the processing site, on-site bioremediation allows the processors in the form of microorganisms to be brought to the contaminated site.

5 HOW DO THE MICROBES DO IT?

"Microbial species are enormously abundant with upwards of a thousand different species found per square yard of land surface. A single species may encompass from 5,000 to 20,000 genetically different strains or varieties, varying in their adaptive qualities and ecological requirements. Microbes are also ubiquitous, found in every terrestrial environment and habitat."[9] In order to sustain life, these organisms rely on catabolic genes to direct the production of enzymes which in turn metabolize available food sources. Although catabolic microorganisms are usually isolated from contaminated sites, it is possible to find a huge metabolic diversity among the many species in a typical soil sample. Not only will these soil bacteria have pathways for the degradation of plant and animal debris, but they will also have some pathways that build elaborate molecules like antibiotics and still other pathways to break down the foreign molecules they encounter. It is often found that complex compounds are broken down into common biochemical intermediates and that the pathways for the catabolism of these intermediates are very common in bacteria. The scientific literature is replete with examples of microbial species with useful catabolic properties. Although some species are encountered more frequently in these searches (due in part to the selection techniques), it is also true that many different genera are reported in these studies.

Catabolic genes are often found on mobile genetic elements such as transposons, which can move to new positions on the bacterial chromosome, and plasmids, which can be transferred between species. It is generally believed that these catabolic genes result from subtle mutations in natural genetic pathways. These mutations enable key enzymes to accept a wider or altered substrate range so that a synthetic compound might be metabolized. For example, naphthalene is a compound that is rare in nature and is present mostly as a by-product of fossil fuel combustion. Yet naphthalene degraders are not difficult to isolate. It is possible that the naphthalene degradation genes are derived from catabolic pathways for structurally related compounds.

If bacteria possess the appropriate catabolic genes, why are measurable levels of the substrate pollutants still present at sites? Often the local environment is lacking in some important nutrient or growth factor. For example, when aromatic compounds must be degraded, the pathways usually include an oxygenation step to break the aromatic ring. The lack of oxygen or necessary nutrients can stop further degradation. In land treatment, fertilizer is added to the contaminated soil and the soil is repeatedly tilled to aerate it thoroughly. In theory, all nutrients needed are made available through this method and the indigenous microorganisms are able to break down the contaminants. Yet it may be difficult to confirm that biodegradation was the mediator for the disappearance of the pollutants. In practice, volatile contaminants may evaporate from the soil as it is tilled.

6 CASE HISTORIES: SUCCESSFUL APPLICATIONS OF NONRECOMBINANT STRAINS

In 1989, the Applied BioTreatment Association (ABTA), the trade association representing bioremediation firms, published its first case history compendium. A total of 47 chemicals and waste products were identified by the industry as having been successfully biodegraded, and bioremediation at 22 sites was reported. One of the ABTA compendium examples described an *in situ* cleanup of a radar early warning station built on a steep slope underlain by a subsurface shale stratum. Black fuel oil (which had been trapped in the shale backfill) oozed towards the ocean. After replacing leaking underground oil transfer lines, a leaching system was installed which sprayed and injected water to force oil out of the shale pit beneath the radar station's concrete pad. Without disrupting the ecologically sensitive site, the oil-water mixture was then captured, treated using submerged, fixed film biotreatment reactors, and reinjected.[10]

Four major environmental biotechnology symposia held between 1987 and 1991 set the stage for numerous case history presentations.[11] At the 1990 Center for Environmental Biotechnology (CEB) conference, Marios Tsezos (McMaster University, Hamilton, Ontario, Canada) and Ronald G.L. McCready (Biotechnology Section, Energy Mines and Resources Canada, Ottawa, Ontario, Canada) described the continuous extraction of radionuclides by a pilot plant utilizing bioabsorption.[12] Bioabsorption is the sequestering of metal ions from aquatic solutions by microbial biomass. It can be used for water pollution control applications, the recovery of metals from dilute complex aqueous solutions, and the extraction of radionuclides.

Today bioremediation case histories can be found in a growing number of newsletter and trade journal articles. In 1990, as reported by *Microbial Cleanup* (May 1992), an accidental opening in a storage tank in southern England caused the diesel pollution of a neighboring property. After physical cleanup, diesel residues still contaminated the soils. Because the polluted area has preservation-order-protected trees and nearby buildings, excavation and landfilling of the contamination was not an option. The treatment of choice, therefore, was *in situ* bioremediation. Addition of air and nutrients through boreholes was done for 6 months to maintain an aerobic environment for the indigenous microorganisms, resulting in an average decrease from 7446 to 280 mg diesel per kilogram soil.

In the October 15, 1992 issue of *Genetic Engineering News*, PCB work by General Electric (Schenectady, NY) and OHM Remediation Services (Findlay, OH) was described. Six sealed caissons were lowered onto contaminated sediments in the upper Hudson River (New York State). Because these polluted sediments had already been dechlorinated anaerobically through indigenous microbial actions, the field work focused on aerobic biodegradation of the dechlorinated or less substituted PCB congeners. Approximately 50% biodegradation was demonstrated over an 11-week experiment. Only indigenous microorganisms were utilized. Addition of nutrients and oxygen improved the system's performance.

Indigenous microorganisms have been utilized in an ambitious project at the Westinghouse Savannah River Laboratory, near Aiken, SC. The site is contaminated with trichloroethylene (TCE), which is polluting the water supply. The TCE is present as a concentrated lens above the groundwater and is responsible for the contamination of an entire groundwater supply. It is not itself a carbon and energy source for any known bacterial species, although certain types of bacteria can break TCE down through a fortuitous catabolism. Some toluene oxidizers, propane oxidizers, and methane oxidizers (methylotrophs) can degrade TCE using the same enzyme that oxygenates their carbon substrate. After this initial step, complete degradation of TCE occurs quickly.

At the Savannah River site, the subsurface soil is known to contain a substantial methylotrophic population. To stimulate the production of more of the right enzyme by this indigenous population, large amounts of methane and air are being pumped into the ground through specially constructed wells. The procedure appears to be increasing the indigenous methylotrophs, both in number and enzymatic activity. The next challenge for this large field project is to demonstrate that enzymatic action is responsible for the degradation of TCE. The increased gas flow through the soil is stripping the volatile TCE from the lens, and this air-stripping accounts for more loss of TCE than the bioremediation. Work is currently underway to either eliminate the TCE as it emerges from the wellheads or optimize degradation conditions in the subsurface.

The French Limited site outside of Houston, TX illustrates a successful application of bioremediation to lagoon cleanup. Just 40 years ago this site was a sand mine, located in an old streambed; 20 years ago the property was purchased by French Limited and licensed by the state as a disposal site. The site was closed in 1971 and listed on the National Priority List 12 years later (1983). Approximately 200,000 cubic yards of organic sludges and contaminated soils were located at the site. Significant decreases in indicator compounds (PCBs, vinyl chloride, benzene, arsenic, and benzo(a)pyrene) were documented, with concentrations of selected indicator chemicals decreasing from 100 ppm to 10 ppm between project days 100 and 300. Details on this case history can be found in *Biotreatment News* (December 1992).

In addition to *in situ* soil, wastewater, and sediment bioremediation technologies, biofilters and bioreactors are also used commercially. The former, first developed in the Netherlands, utilizes bacteria and other microorganisms, usually suspended on natural packing material, to treat waste gases. The market for biotreatment of vapor-phase contaminants has been greatly enhanced with the evolving regulation of toxic gas emissions under the U.S. Clean Air Act. Envirogen, Inc. (Lawrenceville, NJ) has successfully completed the first in a series of field trials utilizing a proprietary biocatalyst-reactor system relying on a pure culture of TCE-degrading microorganisms. The bioreactor was installed at a New York site as a slip-stream biotreatment system on an existing air-stripper which is treating TCE-contaminated groundwater. An average of 90% degradation of TCE-contaminated air was reported by Envirogen.

Despite the above examples, it must be acknowledged that bioremediation is a technology in its infancy, that its accomplishments to date have been limited, and that formidable problems remain before general implementation can occur. For example, suitable catabolic organisms might not be present at a contaminated site. Site characterization to determine whether *in situ* remediation will be effective may add greatly to the cost. The engineering obstacles inherent in agitating large quantities of soil (to add nutrients or oxygen, or to ensure even distribution of contaminants) can be significant. In addition, natural microbial populations might not degrade some compounds completely, which could make the problem worse. This is seen when TCE-degrading organisms produce vinyl chloride, a more dangerous compound. To some extent, these limitations can be overcome by using genetically engineered microorganisms (GEMs).[13]

In summary, over 125 ongoing biology-based remedial actions have been reported in the USEPA's December 1992 *Bioremediation in the Field*, a newsletter produced by the Agency's Offices of Research and Development and of Solid Waste and Emergency Response. Not a single one of these actions is utilizing a genetically engineered microorganism.

7   WHY GENETICALLY ENGINEERED MICROORGANISMS FOR WASTE MANAGEMENT?

As one can see from the case histories, natural microbial isolates from contaminated sites demonstrate the potential for remediation of hazardous waste, either *in situ* or in a specially constructed facility. But natural isolates will never be the whole solution to hazardous waste. Many of the acknowledged limitations might be overcome by genetic engineering of the microbial strains and subsequent introduction of the GEM to a specific site. Several advantages of using GEMs can be cited.[14] First, there are some compounds that defy attempts to degrade them. There might not be pathways in nature that are closely enough related to permit degradation. After all, many synthetic compounds (especially polymers) were created because of their great resistance to breakdown. Second, the majority of waste sites do not have one or a few wastes associated with them; they have a considerable variety of compounds that must be degraded. Microbes which degrade large numbers of compounds have been described, but may not be suitable for every site. Third, natural isolates might degrade these compounds too slowly to be commercially useful. Finally, application of microbes to a site, even a site altered by contamination, presents unique problems. Survival of the microbes is difficult to ensure, for reasons that are not completely clear.[15] It would be useful to give these bacteria a survival advantage before application. When *in situ* problems must be approached, it is sometimes necessary to deliver the catabolic bacteria to a remote site. Transport of the bacteria may be easier if they are genetically manipulated to bind to soil particles whenever they detect contaminants. Programmed cell death after the destruction of the contaminants has also been suggested.[16]

In addition, GEMs can provide critical information about the site, about the composition of the contaminants, and about the microbial response to those contaminants. At a site polluted with organic compounds, three elements must be present for successful bioremediation. First, microorganisms capable of producing catabolic enzymes are necessary. Second, even if microbes have the appropriate genetic code for enzyme production, induction of these catabolic genes is required for degradative activity. Third, the pollutants must also be accessible to the microorganisms' enzymes. Bioreporter bacterial strains can be constructed that provide this critical information.[17] These strains provide a quick and reliable assay for the functioning of the catabolic genes or can be used as indices of the health of the bacterial strains. This type of information can be used to determine whether toxic chemicals are killing the bacteria, whether nutrient limitation is inhibiting bacterial activity against contaminants, or whether the contaminant of interest is physically available to the bacterial cell. This last point is particularly important, since some chemicals may associate with soil particles and resist microbial uptake.[18]

Insert 4 summarizes the advantages of GEM utilization for waste management and cleanup.

INSERT 4

Uses of GEMs

- to improve the biochemical performance and versatility of microbial strains involved in biodegradation
- to develop microbial strains or processes that can be used to evaluate and control environmental performance of engineered biotreatment systems
- to improve the hardiness or staying power of microorganisms in the face of environmental stresses and toxins
- to develop cellular control or containment in order to limit the persistence or spread of organisms used in biotreatment/bioremediation
- to provide important site characterization information

Source: Sayler, G.S. and S.M. Day, SPECIAL REPORT: Bioremediation — experts explore various biological approaches to cleanup, *HAZMAT WORLD,* January 1991.

For these reasons the use of GEMs has been suggested to augment bioremediation processes. GEMs may increase the efficiency of biodegradation, enable added microorganisms to survive at sites, and extend the substrate range of microorganisms. The first step in release of recombinant reporter strains has already been performed.[19] This was a necessary prelude to the release of commercial strains for biomonitoring and biodegradation. With the successful completion of this work, it is likely that field release of GEMs will occur more frequently.

8   REGULATIONS: GLOBAL REGULATORY TRENDS

8.1   Introduction

Waste management and remediation technologies are governed by regulations and government policies. Environmental programs which require the management of wastes at the point of generation and the restoration of contaminated soils and groundwater from poor waste disposal practices create markets for waste treatment methodologies, including those that are biology-based. These programs often rely on permits or compliance orders which specify operational procedures or expected results (cleanup standards) for handling site-specific contaminants.[20,21] Of particular interest to the field applications of biotreatment are the residue, effluent, or emissions levels (acceptable level of cleanliness) specified under each environmental program. Biological treatment may meet risk-based cleanup standards, but not incineration-based levels.

Waste treatment methodologies which rely on microorganisms may also have the microbes themselves separately regulated.[22] Global environmental regulations as they apply to the microorganisms and their use in biology-based waste management applications have developed along two very different tracks. In Europe, Japan, and the United States, microorganisms which have undergone the process of genetic manipulation by man are reviewed separately from those which have been initially isolated from the environment.[23] This is not the case in the evolving Canadian regulations, where similar information is required for naturally occurring and genetically manipulated microbes. In October 1992, Environment Canada circulated a draft product rule which covered all microorganisms regardless of whether they had been genetically manipulated by man. This shift to an expanded universe promises to have a significant impact on bioremediation.

Physically contained microorganisms, such as those growing in bioreactors, regardless of whether they have been genetically engineered, are regulated worldwide to a lesser extent than microbes applied directly to the environment. Yet variation also occurs for this application of environmental biotechnology. Under today's draft U.S. Toxic Substances Control Act (TSCA) regulations, GEMs in a bioreactor are not reviewed by the U.S. Environmental Protection Agency, while the Council of Environmental Communities' (EC) 1990 Council Directive on the contained use of genetically modified microorganisms spells out specific information requirements for low-risk and high-risk microbes, with an emphasis on rapid notification of neighboring countries if an accidental release occurs.[24]

Laboratory research, except in the most recent draft U.S. rules, is addressed by biosafety guidances and internal committees (institutional controls) rather than specific governmental regulations. In the United States, the National Institutes of Health (NIH) "Guidelines for Research Involving Recombinant DNA Molecules" were published in 1986. Classifying a particular GEM as to the risk and application of specified confinement levels served as the basis for these guidelines. Over time, application of these guidelines expanded to include all federally funded research. In 1991, the NIH retired from review of environmental applications, delegating it to the EPA for bioremediation microbes. "Good Developmental Practices for Small Scale Field Research with Genetically Modified Plants and Microorganisms" was developed by the European Organization for Economic Cooperation and Development as the EC's version of deliberate release guidelines.

Product regulation, in general, consists of a notification requirement and the submission of information designed to assist the regulatory agency with its determination of risk potential resulting from the product's commercial use In most countries, protection of confidential data (CBls, or confidential business information) submitted to the governmental body is also addressed. Once the regulators have reviewed the product's submission and are comfortable with its projected health and environment impact, the microorganism will be included on an approved list [e.g., the U.S.'s TSCA Chemical Substances Inventory and Canada's Domestic Substances List (DSL)] and can be used in the field.

A critical parameter for bioremediation applications is whether the field use of non-GEMs requires regulatory notification. At this juncture, both indigenous and nonindigenous naturally occurring microorganisms are "automatically considered" on the U.S. TSCA Inventory and not referred to in the evolving EC regulations, while only non-GEMs in Canadian commerce during 1987/1988 are grandfathered onto the DSL.

On April 23, 1990, the EC's Council published its directives on the contained use and the deliberate release of genetically modified microorganisms.[25] The latter directive requires notification of a national competent authority before undertaking a release and, with the notification, a technical dossier including risk assessment and proposed packaging and labeling. These directives promise to lessen the regulatory burden imposed upon environmental applications of biotechnology in the 1980s by Germany and Denmark. Moreover, any product approved under the EC directives can be marketed throughout the member nations.

To normalize U.K. regulations with the EC directives, in October 1991 the U.K. Secretary of the Department of Environment, the Minister of Agriculture, Fisheries, and Food, the Secretary of State for Scotland, the Secretary of State for Wales, and the Health and Safety Commission published a consultation paper titled *Genetically Modified Organisms (GMOs): Proposed New Regulations*. According to the paper, regulations to control the safety of GMOs have been in place since 1978 and the currently operative regulations are the Genetic Manipulation Regulations of 1989. These rules require notification of the Health and Safety Executive of genetic modification activities, including proposals to introduce GMOs into the environment. Part VI of the Environmental Protection Act 1990 is the statutory authority for 1991 draft U.K. biotechnology regulations. The purpose of Part IV is the prevention or minimization of any damage to the environment which may arise from the escape or release from human control of genetically modified organisms.

The U.K. Legislative language does not rely on an approved list as does the U.S. and Canadian draft rules; instead, individuals are directed to carry out a risk assessment before importing, acquiring, or releasing GMOs and to notify the Secretary of State of his intention. The statute directs the individual to maintain his/her risk assessment for a prescribed length of time but does not require submission of the information to a review agency. In contrast to the U.S./Canadian listing as signifying approval, the U.K. approach relies on a consent (similar to a permit or compliance order) granted by the Secretary of State.[26] A prohibition order may be issued by the Secretary of State if he is of the opinion that the action would involve a risk of causing damage to the environment. The Secretary of State is directed to maintain a registry of consent applications and approvals.

GMO-specific statutory language is quite common in the European Community. In addition to the British law, a Swiss referendum was passed on May 17, 1992. The enactment of the Swiss Beobachter Initiative introduced genetic technology into the federation's constitution. This vote appears to be the first of a series, with future ones addressing patent issues, bans on transgenic animals, and deliberate releases of genetically modified plants.[27]

The French National Assembly approved on June 26, 1992 the second reading of a law on the release of GMOs into the environment. The proposal institutes a legal framework for monitoring and control of the use and release of GMOs in order to protect human health and the environment. Academic and other research and development institutions were initially subject to public hearings and government approval under this proposal, but the former requirement was dropped and a confidential business information protection added before approval by the Assembly.[28]

Japan's biotechnology regulations generally follow United States and EC requirements. In the Japanese "Points to Consider" four focus areas are emphasized: (1) genetic considerations, (2) ecological considerations, (3) pathological considerations, and (4) considerations for post–release control. Research guidelines, developed by the Japanese Ministry of Education, Science, and Culture, are based on early versions of NIH guidelines, but according to the Congressional Office of Technology Assessment are more stringent. In June 1986, the Ministry of International Trade and Industry (MITI) issued guidelines for industrial applications of GEMs, and the Japanese Environment Agency has drafted safety guidelines for field tests of GEMs.[29] In 1986, the Japanese Bioindustry Association undertook a project titled "Study of Safety Concerns of Recombinant Microorganisms in the Environment" supported by the MITI. The study's objectives were (1) to determine the factors regulating survival and growth of an introduced microorganism, (2) to evaluate the effects of an introduced microorganism on a soil ecosystem, and (3) to assess the equivalence of soil microcosm and field data.[30] *Sphingomonas paucimobilis* SS86, formerly named *Pseudomonas paucimobilis*, although not a recombinant microorganism, was used for this study.

The Organization of American States (OAS) published its *Guidelines for the Release into the Environment of Genetically Modified Organisms* in June 1991. In its prologue, the authors observed that "no biotechnology regulations exist in Latin America and the Caribbean, with the exception of a few research institutes that have established internal biosafety assessment procedures for work with biotechnological techniques. This is not only because of the small research effort currently being undertaken in the region, but also because of the lack of political or public pressure to establish these regulations. But with the rapid advent of commercial live products obtained through biotechnology, it is urgent to establish mechanisms.... The guarantee of standards equal to the ones applied in developed countries is an important objective, so as to maintain the confidence of the scientists and the general public in the new technologies." In its introduction, the guide goes on to say that its recommendations are based on the world's most recent experience, with particular credit going to Canada. Like the draft Canadian rule, the OAS guidelines address all microorganisms regardless of the process by which they are derived.

8.2 Common Requirements

In general, governments want to know the name of the microorganism and its genetic lineage. One common nomenclature for microbes has not been adapted internationally. In the United States, for example, species or strains may be numbered. Whether the microbe has been patented or whether a patent is pending is usually asked. Perhaps because it is the most recently developed, the Canadian draft rule asks for information on consortia, in addition to individual microorganisms.

Data on the site where the microorganism will be utilized are also routinely requested. See Chart 1 for representative site characterization information. It is this recurrent data set which suggests that the risks of the microbial product will be reviewed differently from chemicals and nonmicrobial pesticides sold for environmental applications. It appears that a dual review process (product and process) could occur for every remedial action using microorganisms in Canada and member countries of the

Organization of American States, while only environmental applications of GEM(s) in countries such as U.S., U.K., and Germany would require both microbe approval and applicable environmental permits.

CHART 1

Basic Site and Sediment Characteristic Information Needs

1) characterization and concentration of wastes, particularly organics in the contaminated sediments

2) microorganisms present in the soils, sediment/groundwater and their capability to degrade, cometabolize, or absorb the contaminants

3) biodegradability of waste constituents (half–life, rate constant)

4) biodegradation products

5) depth, profile, and aerial distribution of constituents in the contaminated media

6) soil, sediment, and/or water properties for biological activity (such as pH, oxygen content, moisture and nutrient contents, organic matter, temperature, etc.)

7) soil/sediment texture, water-holding capacity, degree of structure, erosion potential of the soil

8) hydrodynamics of the site

Source: Handbook on *In Situ* Treatment Of Hazardous Waste Contaminated Soils, Risk Reduction Engineering Laboratory, U.S. Environmental Protection Agency.

9   PUBLIC POLICY ISSUES RELATING TO BIOREMEDIATION STILL EVOLVING

According to the Greek philosopher Sophocles, "one must learn by doing the thing. For although you think you know it, you may have no certainty until you try." This quote is quite appropriate to the evolving field of biology-based waste treatment. Today, hundreds of field applications using naturally occurring microorganisms have been completed or are underway. Field experience is rapidly accumulating. The historic association of biotreatment with black box technology is being disproved as successful bioremediation case histories multiply. International information exchange systems promise to contribute to the bioremediation knowledge base worldwide.

Environmental biotechnology is a very sophisticated science, and genetic engineering can provide more solutions to environmental problems than current commercial applications suggest. If GEMs are acknowledged by scientists as a potential asset for bioremediation, they are also viewed with a mixture of distrust and fear by many other groups. The same uncertainties that surround field releases of GEMs for agricultural uses are found for bioremediation.[31] The perceived risks have fostered the growing regulation of the biotechnology industry.[32,32a]

Some public interest groups have chosen to segregate biology-based waste treatment utilizing GEMs from such applications relying on non–GEMs. The following National Wildlife Federation quote sums this up: "The potential environmental impacts of [GEMs] are almost unlimited.... There is no dispute that engineered organisms as a group will possess a higher degree of genetic novelty than traditionally bred or naturally oc-curring organisms and that our environmental experience with such or-ganisms is virtually nil."[33] In 1992 a public interest group called the California Action Council published The *Overselling of Bioremediation: A Primer for Policy Makers and Activists.*[34] In it, the following observation was made: "Considerations of the ecological impacts of bioremediation tech-nologies have only begun to be deliberated. It is uncertain if spreading trillions of non-native microorganisms into a given environment will ad-versely affect it, or if the amendments added to enhance the growth of the microorganisms will have any negative effects on the surrounding ecosys-tem." Yet these voices of caution are relatively few for biology-based waste treatment. Because the use of microbes promises to solve toxic substances problems, the public and the press appear to be more receptive to the technology.

It is anticipated that as the comfort level increases with routine uses of nonengineered microorganisms, future field applications of GEMs, which can bring improvement in the performance currently achievable, may be more readily accepted by the public. But if regulations become onerous at the research level, hindering transition from the laboratory to the field, then environmental biotechnology will be slow to emerge as a waste management option.[35]

In addition to improving our scientific and field–experience founda-tion, political leadership is critical for widespread environmental applica-tions of GEMs. If countries such as Japan or the Netherlands or the United States choose to invest in optimizing biology-based waste treatment per-formance, the field may rapidly grow. In closing, regardless of which countries take the lead, field experience in bioremediation is accumulat-ing. The key question about whether biodegradation and/or biotransfor-mation are magic bullets for inexpensive, permanent waste management and remedial actions may be nearing an answer.

ACKNOWLEDGMENTS

This work was supported in part by the University of Tennessee, Waste Management Research and Education Institute, the U.S. Department of

Energy (DOE), and U.S. Air Force grants DE–F605–91ER61193 and F49620–92–J0147, respectively, and by the Office of Environmental Research, U.S. Department of Energy, under contract DE–AC05840R21400 with Martin Marietta Energy Systems, Inc.

REFERENCES

1. The Roper Organization, 205 East 42nd Street, New York, New York 10017.
2. Netherlands, Ireland, Italy, Germany, Spain, France, United Kingdom, Belgium, Greece, Portugal, Denmark, and Luxembourg.
3. The European Community and Environmental Protection, July 1992.
4. U.S. Environmental Protection Agency (USEPA), Reclamation and Redevelopment of Contaminated Land, Volume II, European Case Studies, EPA/600/R-92/031, USEPA, March 1992.
5. Environmental Policy in the European Community Periodical 5, 1990.
6. *WasteTech News*, Volume 4, Number 2, September 23, 1991.
7. User Manual, VISITT (Vendor Information System for Innovative Technologies), Version 1.0, EPA/542/R-92/001, Office of Solid Waste and Emergency Response, U.S. Environmental Protection Agency, June 1992.
8. Understanding Bioremediation: A Guidebook for Citizens, U.S. Environmental Protection Agency, February 1991.
9. U.S. Congress, Office of Technology Assessment (OTA), New Developments in Biotechnology: Field-Testing Engineered Organisms; Genetic and Ecological Issues, OTA–BA–350, OTA, May 1988.
10. Applied BioTreatment Association (ABTA), Case History Compendium, ABTA, Washington, D.C., November 1992, p. 23.
11. Environmental Biotechnology: Reducing Risks from Environmental Chemicals through Biotechnology, symposium hosted by the University of Washington, July 1987; HMCRI's Biotreatment Meeting, December 1989; Flask to the Field Conference, hosted by the Center for Environmental Biotechnology, University of Tennessee, October 1990; and Battelle's In Situ and On Site Bioreclamation, April 1991.
12. Tsezos, M. and R.G.L. McCready, The pilot plant testing of the continuous extraction of radionuclides using immobilized biomass, in *Environmental Biotechnolog for Waste Treatment*, G.S. Sayler, R. Fox, and J.W. Blackburn (Eds.), Plenum Press, New York, 1991, pp. 249–260.
13. Definitions for genetically engineered microorganisms (GEMs) and genetically modified microorganisms (GMOs) vary from country to country. All countries but the U.S. utilize the term GMO. In this chapter, genetically modified microorganisms will be the expression of choice, but when talking about U.S. researchers and regulators, the term GEM is more appropriate.
14. Lindow, S.E, N.J. Panopoulos, and B.L. McFarland. Genetic engineering of bacteria from managed and natural habitats, *Science* 244:1300–1307, 1989.
15. LeJoie, C.A., S.Y. Chen, K.C. Oh, and P.F. Strom, Development and use of field application vectors to express nonadaptive foreign genes in competitive environments, *Appl. Environ. Microbiol.* 58(2): 1992.
16. Bej, A.K., M.H. Perlin, and R.M. Atlas, Model suicide vector for containment of genetically engineered microorganisms, *Appl. Environ. Microbiol.* 54: 2472–2477, 1988.
17. Burlage, R.W., A. Heitzer and G.S. Sayler, Bioluminescence: a versatile bioreporter for monitoring bacterial activity, *BFE* 9(11/12):704–709, 1992.

18. Atlas, R.M., G. Sayler, R.S. Burlage, and A.K. Bej, Overview: molecular approaches for environmental monitoring of microorganisms, *BioTechniques* 12(5):706–717, 1992.

19. Shaw, J.J., F. Dane, D. Geiger, and J.W. Kloepper, Use of bioluminescence for detection of genetically engineered microorganisms released into the environment, *Appl. Environ. Microbiol.* 58:267–273, 1992.

20. In addition to market creation, the recent complexity of the United States' hazardous waste management rules has limited the structural design of bioremediation units.

21. Day, S.M., Regulatory considerations, in *Biotechnology for Hazardous Waste Treatment*, D.L. Stoner (Ed.), Lewis Publishers, Boca Raton, FL, in press.

22. It is important to note that environmental biotechnology product regulations in most countries are not final and at the time of publication are still evolving.

23. Although the U.S. principles for regulatory review state that federal government regulatory oversight should focus on characteristics and risks of the biotechnology product, not on the process by which it is created, the draft regulatory structure effectively exempts non-GEMs from review.

24. Glass, D., Obtaining regulatory approval and public acceptance for bioremediation projects using genetically engineered organisms, in the Proceedings of the 1993 International Symposium on *In Situ* and On-Site Bioreclamation, in press.

25. Council Directive of 23 April 1991 on the contained use of genetically modified microorganisms. Official Journal of the European Communities 8.5.90 No. L117/1 and Council Directive of 23 April 1990 on the deliberate release into the environment of genetically modified organisms. Official Journal of the European Communities 8.5.90 No. L117/15.

26. A similar approach has been suggested for approval of field tests in Russia. V.G. Dbabov, "Introduction of Genetically-Altered Microorganisms into the Environment (Review)" All-Union Scientific Research Institute of Genetics and Selection of Industrial Microorganisms, Moscow. Translated from *Mikrobiologiya*, 60(1):5–9, 1991.

27. Hodgson, J., Swiss pass biotech referendum, *Bio/Technology* 10:627, June 1992.

28. French Parliament Approves Proposal on Genetically Modified Organisms, The Bureau of National Affairs, June 29, 1992.

29. Biotechnology in a Global Environment. Congress of the United States, Office of Technology Assessment, OTA-BA-494, October 1991, p. 192.

30. Japan Bioindustry Association (JBA), Study on Safety of Recombinant Microorganisms in the Environment, JBA, 1992.

31. Lindow, S.E. and N.J. Panopoulos, Field tests of recombinant ice — *Pseudomonas syringae* for biological frost control in potato, in *Release of Genetically Engineered Microorganisms*, M. Sussman, C.H. Collins, F.A. Skinner, and E.E. Stewart-Tull (Eds.), Academic Press, San Diego, CA, 1988.

32. Kolata, G., How safe are engineered organisms?, *Science* 229:34–35, 1985.

32a. Marx, J.L., Assessing the risks of microbial release, *Science* 237:1413–1417, 1987.

33. Mellon, M., Biotechnology and the Environment, National Biotechnology Policy Center, National Wildlife Federation, ISBN 0-912186-99-2.

34. Stabinsky, D., *The Overselling of Bioremediation: A Primer for Policy Makers and Activists*, Caliornia Action Council, 1992.

35. For example, the German biotechnology rule does not allow importation of microorganisms tor research purposes without government approval. The 1991 draft TSCA rule requires university laboratories to notify the Agency before undertaking field work using GEMs.

Chapter 5

INTRODUCTION OF GENETICALLY ENGINEERED MICROORGANISMS INTO THE ENVIRONMENT: REVIEW UNDER USDA, APHIS REGULATORY AUTHORITY*

Arnold S. Foudin and Cyril G. Gay

TABLE OF CONTENTS

* Disclaimer: This article is not purported to be the official version of the rules and regulations administered by USDA, APHIS.

1   INTRODUCTION

The United States Department of Agriculture (USDA) Animal and Plant Health Inspection Service (APHIS) regulates the introduction of genetically modified organisms which are or may be derived from plant pests or which are used in veterinary products such as vaccines. APHIS has established a division called Biotechnology, Biologics, and Environmental Protection (BBEP) which oversees the regulatory process and provides a "circle of protection" to safeguard American agriculture. To ensure the safe use of genetically modified organisms, the "circle of protection" includes movement and environmental introduction permitting programs, environmental impact assessments, product licensing, public access to information, and liaisons with other federal agencies which oversee biotechnology programs. In addition, BBEP supports nonbiotechnology assessment and monitoring activities related to plant protection and production of veterinary biologics.

APHIS was the first federal agency to promulgate a set of codified rules governing the introduction of genetically engineered plants and microorganisms. As a result, more than 40 safe field tests of genetically modified microorganisms have been conducted in the United States. APHIS has also evaluated over 400 field tests of genetically engineered plants. Accordingly, APHIS has acquired considerable experience with the scientific, environmental, and public health issues related to the deployment of genetically engineered plants and microorganisms in the environment. This experience has contributed to the development of flexible regulatory review procedures that can accommodate changing technologies, ease the regulatory burden, and at the same time address public concerns regarding the safety of biotechnology products.

2   LAWS AND REGULATIONS

The authority of the various federal agencies to regulate biotechnology was laid out in the "Coordinated Framework for Regulation of Biotechnology", which was published in the *Federal Register*, Vol. 51, No. 123, 23302–23350, June 26, 1986. This document defined the responsibilities of the Department of Agriculture, the Food and Drug Administration, and the Environmental Protection Agency and set the policy for subsequent regulations. The policy stated that products developed by the methods of biotechnology do not differ fundamentally from conventional products and that the existing regulatory framework is adequate to regulate biotechnology. Thus, existing statutes have provided the basis for agency jurisdiction over research activities and the introduction of genetically engineered organisms into the environment. The USDA's responsibility to protect American agriculture is divided into two broad categories — plant protection and animal health.

2.1   Regulation of Plant Pests

Operating under the "Coordinated Framework for Regulation of Biotechnology", APHIS promulgated regulations governing the safe introduction of genetically engineered plants and microorganisms into the environment, which were published in the *Federal Register* on June 16, 1987 (52 FR 22892–22915). The biotechnology regulations were codified in Title 7 of the Code of Federal Regulations (CFR), Part 340. The biotechnology regulations set forth in 7 CFR 340 are based on the USDA's statutory authority under the Federal Plant Pest Act of 1957 and the Plant Quarantine Act of 1913. To be regulated under these statutes, a genetically engineered organism that is intended to be introduced into the environment must be a plant pest, or derived from a plant pest, or believed to be a plant pest, and it appears on the list of regulated articles in 7 CFR 340. "Plant pest" is defined as "any living stage of any insects, mites, nematodes, slugs, snails, protozoa, or other invertebrates, bacteria, fungi, other parasitic plants or reproductive parts thereof, viruses, or any other organisms similar to or allied with any of the foregoing, or any infectious substances of the aforementioned which are not genetically engineered, and which can directly or indirectly cause disease or damage in any plants or parts thereof, or any processed, manufactured, or other products of plants".

2.2 Regulation of Veterinary Biologics

The authority for the regulation of veterinary biologics, such as genetically engineered vaccines used for animal health, is provided in the Virus, Serum, Toxin Act (VSTA) of 1913. VSTA was amended by the Food Security Act of 1985 to authorize APHIS to exercise regulatory authority over the production, distribution, and evaluation of all veterinary biologics. The USDA was also mandated to promulgate regulations and procedures consistent with the Act to ensure that veterinary biologics are pure, safe, potent, and efficacious. The pursuant regulations were codified in 9 CFR, Parts 101 to 118. The regulations in 9 CFR 101.2 (w) define veterinary biological products to be "all viruses, serums, toxins, and analogous products of natural or synthetic origin, such as diagnostics, antitoxins, vaccines, live microorganisms, killed microorganisms, and the antigenic or immunizing components of microorganisms intended for use in the diagnosis, treatment, or prevention of diseases of animals". The definition intentionally excludes products that are animal drugs (e.g., antibiotics and analgesics).

2.3 Regulations under NEPA

On receipt of a request to conduct a field test with a regulated genetically engineered organism, APHIS must determine whether the provisions of the National Environmental Policy Act (NEPA, 1969) are triggered. NEPA is invoked whenever there is a proposed federal action with a potential for significant impact upon the human environment. An APHIS intent to approve the introduction of a regulated article into the environment is such a federal action.

The Council on Environmental Quality regulations under 40 CFR, Parts 1500 to 1509, implement NEPA (CFR Title 7, Part 1b). According to these regulations, actions by a government agency such as APHIS may fall into one of four categories: (1) actions which are exempt due to categorical exclusion, (2) actions which are covered by an existing environmental document, (3) actions which require preparation of an Environmental Assessment, or (4) actions which require preparation of an Environmental Impact Statement. Accordingly, not every federal action is subject to the same scrutiny. Obviously, certain types of actions may have more significant environmental effects than others. NEPA regulations are intended to facilitate the federal decision-making process by requiring an analysis of alternatives to the proposed federal action, a comprehensive consideration of the environmental effects of the proposed activity, and involvement of the public in the decision-making process.

3   THE REVIEW PROCESS

Procedures for the regulatory oversight of organisms or products derived from biotechnology have been established within APHIS to implement the regulatory framework described in the previous section.[1-3] The cornerstone of these procedures is a review process to ensure protection of the human environment through proper scientific analysis, peer review (veterinary biologics), public notification, and documentation of the decision-making process.

The review process is based on the principles of risk analysis. This provides a mechanism to support decisions on the use of genetically engineered microorganisms and experimental vaccines in the field. Risks to the target host animal species, public health, and the environment are assessed. The objective is to link science to decision-making. That basic function is the same whether risks to animals, people, or the environment are being considered.

One important feature of the methodology used by APHIS is that significant effort is placed on identifying hazards early in the risk analysis process. Hence, risks are identified based on the molecular and biological properties of the microorganism being evaluated; risks are not assumed to exist just because the microorganism has been derived through the process of biotechnology. This approach is consistent with accepted standards for conducting risk analysis.[4,5] The multifactorial approach (i.e., the inclusion of a qualitative risk assessment for animal and human health and for environmental safety) is also consistent with recent recommendations by the National Research Council.[6]

The BBEP review procedures are documented in the Environmental Assessment that is prepared according to the provisions of NEPA. It is a succinct document that briefly provides sufficient evidence and analysis for determining whether or not to prepare an Environmental Impact Statement or Finding of No Significant Impact (40 CFR, Part 1508.9). Each Environmental Assessment contains a brief discussion of the need for the proposed action; a description of departmental regulations; the conditions under which authorization to conduct a field test is approved or denied; the safety characteristics of the genetically engineered organism; the potential environmental impacts of the field test; alternative actions, as required by Section 102(2)(3) of NEPA; and a list of federal agencies and persons consulted.

3.1   Review of Genetically Engineered Microorganisms under 7 CRF 340

In accordance with regulations of 7 CFR 340 governing plant pests, permits are required for the introduction (importation, interstate movement, or release into the environment) of regulated articles such as pathogenic bacteria. These permits are issued by Biotechnology Permits, BBEP, subsequent to an established review process, which included the preparation of an Environmental Assessment as described above. The finding noted in the Environmental Assessment that is essential for the issuance of a permit is that the genetically engineered organism in question does not present a risk of introducing or disseminating a plant pest into the environment. The following critical points are examined in detail with regard to genetically altered bacteria:

1. the scientific nomenclature and taxonomy of the bacterium
2. the general biological properties of the bacterium, its habitat, and the multiplication rate of the wild type/parental strain
3. the genotypic and phenotypic characteristics of the altered microorganism, particularly whether it has any intrinsic plant pathogenic properties, whether any trait has been altered, or whether it has acquired any new trait that makes it a new plant pest
4. determining whether the host range of the genetically engineered microbe has been altered, particularly when the parental wild type is a plant pathogen
5. stability of the genetic construct, either plasmid borne or chromosome borne, and the potential for gene transfer
6. the nature of the marker gene(s) (antibiotic or nonantibiotic)
7. selection pressure for the maintenance of the strain
8. the physical environment into which the organism is being introduced
9. the potential impacts to the environment and its components due to the introduction of the genetically engineered microorganism; assessment of effects on nontarget organisms, endangered flora and fauna, and human and animal health; examination of the field plot design to evaluate the kind of physical and biological containment features built into them; and, more importantly, critical examination of the mitigation procedures that are in place should an unexpected event occur under field conditions
10. field test monitoring protocols, including the adequacy of sampling methods for the detection of nonculturable bacteria
11. review of field data reports from any previous field experiments

Normally, the Environmental Assessment will document a Finding of No Significant Impact. However, if the alternative conclusion is reached, it may be necessary to initiate an Environmental Impact Study or to reject the permit application. To date, no application has required an Environmental Impact Statement or has been rejected because the applicant is given the opportunity to withdraw the application at any time during the review process. In some cases, it may be necessary for the applicant to modify the protocol in order to address any concerns raised by the reviewer. Post-permitting amendments are allowable as long as the changes do not affect the organism's plant pest status or the Finding of No Significant Impact in the Environmental Assessment.

3.2 Review of Veterinary Vaccines

The regulatory oversight of all veterinary vaccines resides within Veterinary Biologics, BBEP. It should be emphasized that the review process for vaccines derived from biotechnology is not fundamentally different from the types of reviews that have traditionally occurred for conventional live veterinary vaccines (e.g., anthrax, rabies, *Bordetella bronchiseptica*, etc.). The primary difference is that specific questions relative to the molecular and biological properties of genetically engineered microorganisms have been identified so as to provide a structured review and analysis of proposed contained or environmental releases. The risk analysis method developed by Veterinary Biologics (referred to as Biotechnology Analysis) (Table 1) is flexible. Because of the nature and use of veterinary biologics in the field, it is a multifactorial approach to risk assessment.

Specific information on the molecular and biological properties of the experimental vaccine is submitted to Veterinary Biologics through the

TABLE 1

Veterinary Biologics Biotechnology Analysis

I. Objectives/proposal

II. Characterization of the genetically altered microorganism
   A. Molecular properties
   B. Biological properties

III. Risk Assessment
   A. Hazard identification
      1. Animal safety
      2. Public health safety
      3. Environmental safety
   B. Release assessment
   C. Risk characterization

IV. Risk management
   A. Contained release
   B. Environmental release
      1. Monitoring
      2. Mitigative procedures

completion of a Summary Information Format (SIF). For live vaccines, separate SIFs are available for gene-deleted vaccines and recombinant vector vaccines. The SIF is supplied to the applicant on a computer diskette with instructions for completion to streamline the review process. Firms are encouraged to complete the SIF in the early stages of product development.

Upon receipt of a completed SIF, Veterinary Biologics initiates the risk analysis process (i.e., the preparation of a Veterinary Biologics Biotechnology Analysis). The objective is to describe explicitly the methods used to identify and assess potential risks and to present results in a form useful for decision-making when risks are identified.

Since the molecular properties of genetically engineered vaccines are often considered proprietary information by the applicant, the Biotechnology Analysis conducted by Veterinary Biologics is peer reviewed outside of the BBEP only by scientists who have signed a confidentiality statement.

The decision-making process is documented in the form of either a Decision Document or an Environmental Assessment. Both documents summarize the Veterinary Biologics Biotechnology Analysis and, thus, the rationale leading to the decision to approve or deny a field test. The type of document prepared is based upon whether or not hazards are identified in the risk analysis. When no hazards have been identified, a Decision Document is prepared and filed with the product license application. The Environmental Assessment is prepared in accordance with NEPA when potential hazards to the human environment are identified.

4   APPLICATION PROCESS

4.1   Field Release Permits Issued under 7 CFR 340

Any person applying for a permit for the introduction of a regulated article must be an American citizen. The process for obtaining a permit for field release is separate from that for movement (i.e., shipment between contained facilities). The permitting process for movement, which must be completed within 60 days, will not be described here.

The permitting process for field releases must be completed within 120 days and begins with the receipt of an application which is checked for completeness. The public is notified of the receipt and availability of the application in the *Federal Register*. The application, along with a preliminary assessment, is sent to the cognizant state for comment within the first 30 days. The application is reviewed as described in the previous section, and an Environmental Assessment is prepared. The public is again notified in the *Federal Register* of the availability of the Environmental Assessment document.

After the permit application has been reviewed with appropriate documentation, a permit for conducting the proposed field experiment with a genetically engineered organism is issued to the applicant. The permit will contain both standard and supplemental permit conditions. Supplementary conditions typically ask the applicant to follow certain case-specific conditions while conducting the field experiments. Sometimes it also contains certain plant quarantine conditions imposed by the regulatory authorities of the state in which the experiment will be conducted. Copies of the permit are sent to state regulatory authorities and the APHIS, PPQ Regional Biotechnologists, who conduct on-site inspections. The field site is inspected before the start of the experiment, and after the termination of the experiment. APHIS also reserves the right to inspect the field test site without notice while the experiment is ongoing to verify compliance with the permit conditions. The field test data are analyzed for the persistence of the introduced organism, for migration both horizontally and vertically, and for any gene transfer to nontarget organisms. This analysis is conducted to advise the applicant in planning future experiments, if the situation so warrants.

APHIS also requires a field data report within 1 year after the termination of the experiment. These reports provide important information about any unforeseen changes in the biological behavior of the organism, such as pleiomorphic effects or site-specific mutagenesis caused by the insertion of foreign DNA. The information from these field data reports has been of value in shaping future biotechnology policy and regulations.

Of the more than 400 permits for field tests issued to date, 11 permits were for genetically engineered bacteria (Table 2). In response to potential applications, BBEP, APHIS also has written opinion letters to the effect that certain genetically modified bacteria are not regulated articles. These are *Rhizobium meliloti*, *R. leguminosarum* bv. *viciae* and *trifolii*, *Bradyrhizobium japonicum*, and *Agrobacterium radiobacter*.

4.2   Shipment of Veterinary Biologics under 9 CFR 103.3

Prior authorization from APHIS is required to ship by any means unlicensed experimental biological products anywhere into or out of the United States. This is consistent with the intent of the VSTA, which was enacted by Congress to prevent the importation and shipment of worthless, contaminated, dangerous, or harmful veterinary biological products. Specific requirements have been established by APHIS for approving the shipment of live genetically engineered vaccines for the purpose of evaluating such products either in containment or in the field.[7]

Authorization is required from the Administrator of APHIS in accordance with the provisions of 9 CFR 103.3 to ship experimental veterinary biologics for the purpose of gathering data in support of a product license.

TABLE 2

Field Tests of Genetically Modified Bacteria in the United States

Modified Strains	Genetic Trait	Location/Year	Researcher
Rhizobium meliloti	Increased N fixation	Wisconsin 1987	Bio Technica
R. meliloti	Kan., Strep.	Wisconsin 1988	Bio Technica
Bradyrhizobium japonicum BJB100/BJB2000	Kan., Strep. npt II	Wisconsin Iowa 1988	Bio Technica
R. meliloti RMBPC-2	Increased N fixation, Kan., Strep spec.	Wisconsin 1992	Research Seeds
R. leguminosarum bv. *trifolii* and bv. *viciae*	Improved competition	Wisconsin 1989	Eric Triplett
Xanthomonas campestrics pv. *campestris*	Luciferase	Alabama 1990–92	Auburn University
Pseudomonas syringae pv. *syringae*	Tn5 marker	Wisconsin	University of Wisconsin
Clavibacter xyli	δ-endotoxin *Bacillus thuringiensis*	Maryland Illinois Minnesota Nebraska 1987, 1989–1992	Crops Genetics International
Pseudomonas spp.	Luciferase	Alabama 1991	Auburn University

Note: Abbreviations — Kan., kanamycin resistance; Strep., streptomycin resistance; Tn5, avirulent transposon mutation.

Requests for authorization to ship experimental biological products for contained or environmental release studies must contain the following:

1. One copy of a permit or letter of permission from the proper state or foreign country involved.
2. Two copies of a tentative list of the names of the proposed recipients and quantity of experimental product that is to be shipped to each individual. In the event of subsequent changes, additional information is to be furnished when such facts are known.

3. Two copies of a description of the product, recommendations for use, and results of preliminary research work (provided in a completed Summary Information Format).

4. Three copies of labels or label sketches which show the name of the product and bear a statement, "Notice! For Experimental Use Only — Not for Sale", or equivalent. The U.S. Veterinary License legend must not appear on such labels.

5. Two copies of a proposed general plan covering the methods and procedures for evaluating the product and for maintaining records of the quantities of experimental product prepared, shipped, and used. (The plan should include proposed biosafety level assignments for the facility, containment equipment, and operational procedures, as well as certification of these proposed levels by the Institutional Biosafety Committee [IBC]). At the conclusion of field studies, results must be obtained, summarized, and submitted to APHIS.

6. Data acceptable to the Administrator demonstrating that use of the experimental biological product in meat animals is not likely to result in the presence of any unwholesome condition in the edible parts of animals subsequently presented for slaughter.

7. A statement from the research investigator or research sponsor agreeing to furnish, upon the request of VB, additional information concerning each group of meat animals involved prior to movement of these animals from the test site. Such information must include the owner's name and address; the number, species, class, and location of the animals involved; the date the shipment is anticipated; along with the name and address of the consignee, buyer, commission, firm, or abattoir.

8. Any additional information the Administrator may require in order to assess the product's impact on the environment.

APHIS has authorized 27 field tests of live genetically engineered veterinary vaccines in the United States (Table 3). All approved field tests have been for viral vaccines. To date, no field tests have been approved for genetically engineered bacterial vaccines. APHIS has approved the shipment of live genetically engineered vaccines to other countries for the purpose of conducting studies in containment; however, environmental releases or field tests in foreign countries have not yet been approved.

5   DOCUMENTATION AND PUBLIC ACCESS TO FEDERAL DECISION-MAKING INFORMATION

All documents related to applications for the introduction of regulated articles (both plant pests and veterinary biologics) and the review process to approve or deny a field test are public documents and are made available to anyone, subject to the provisions of the Freedom of Information

TABLE 3

Field Tests of Live Genetically Engineered Veterinary Vaccines in the
United States

Product	Genetic Trait	Location/Year	Manufacturer
Pseudorabies vaccine	tk–	Illinois Iowa Minnesota Michigan 1985	TechAmerica
Pseudorabies vaccine	tk– gpX–	Illinois Indiana Iowa Minnesota Missouri Nebraska 1987	Diamond
Pseudorabies vaccine	tk– gpX– beta-gal+	Illinois Iowa Minnesota 1987	SyntroVet
Pseudorabies vaccine	tk– gpIII–	Illinois Indiana Iowa Minnesota Nebraska 1988	TechAmerica
Rabies vaccine	Vaccinia vector gpG+	Virginia 1989	Wistar Institute
Pseudorabies vaccine	tk– gpX– gp1–	Illinois Iowa Indiana Minnesota Nebraska North Carolina 1989	SyntroVet
Rabies vaccine	Vaccinia vector gpG+	Pennsylvania 1991	Wistar Institute
Rabies vaccine	Vaccinia vector gpG+	New Jersey 1992	Jefferson College

Note: Abbreviations — tk–, thymidine kinase gene deleted; gpx–, glycoprotein x gene deleted; gpIII–, glycoprotein III gene deleted; gpl–, glycoprotein 1 gene deleted; beta gal+, beta galactosidase gene insertion; gpG+, glycoprotein G gene insertion.

Act (FOIA). The FOIA, as amended, was enacted by Congress to give the public limited access to government records. By law, information in the possession and control of APHIS that qualifies as a record and that is not exempt from disclosure under the FOIA must be made available upon written request. APHIS must respond to a FOIA request within 10 working days. Confidential business information, as defined by the Act, as well as trade secrets from the private sector are exempt. The public should submit requests for Decision Documents and Environmental Assessments directly to the FOIA Office:

> U.S. Department of Agriculture
> APHIS, Freedom of Information
> 4700 River Road, Unit 50
> Riverdale, Maryland 20737

6 TWO CASE HISTORIES

6.1 Field Test of *Xanthomonas campestris* pv. *Campestris*

On February 15, 1990, APHIS approved a request from Auburn University, Auburn, Alabama to conduct a controlled field test of the plant pathogenic bacterium *Xanthomonas campestris* pv. *campestris*. The field test was conducted on a small test plot on agricultural land in Macon County, Alabama. The bacterium was modified using recombinant DNA techniques to insert the luciferase gene complex from the marine bacterium *Vibrio fischerii* into the *Xanthomonas* chromosomal genome. Under appropriate conditions the recombinant *X. campestris* pv. *campestris* produces light, thus facilitating the ability to monitor the movement and spread of the bacterium in the environment. *Xanthomonas campestris* pv. *campestris* is the causal agent of black rot of crucifers (e.g., cabbage); therefore, it is a regulated article under the provisions of 7 CFR 340. As such, a permit is required to conduct a field test of the bacterium. The specific objective of the proposed field test was to collect information on the usefulness of bioluminescence as a tool to study the etiology of the disease process of this plant-pathogenic bacterium *in situ* under standard agricultural conditions as compared to observations in the greenhouse. Classically, the only methods available to study the disease process were to collect infected material and plate out the organisms or to kill and fix plant material for light and electron microscopic examination. In making the decision to issue a permit to conduct the field test, the following critical areas were examined in detail:

1. the molecular properties of the recipient, donor, and vector organisms; e.g., cloning site and the function of the inserted gene

2. the safety characteristics of the recombinant bacterium; e.g., altered virulence potential, genetic and phenotypic stability, threshold of detection, reversion to wild type

3. the ecological characteristics of the recombinant bacterium; e.g., the host/range specificity, identify of biological vectors, and survivability of the bacterium in the target environment

General concerns raised by this proposal included the pathogenicity of *X. campestris*, its survivability in the environment, and its ability to infect nonhost plants or animals. Each concern was properly addressed by Auburn University with the following studies, mitigative procedures, and monitoring:

1. Laboratory studies showed that the presence of the foreign DNA in the recombinant bacterium decreased its viability and virulence on host plants. The recombinant bacterium had no detectable plasmids and was refractory to attempts to conjugate with plasmids.

2. Plants inoculated with the recombinant bacterium were monitored for any unusual disease symptoms during the test period. Plant parts and soil removed for analysis in the laboratory were transported in containment and were properly disposed of in the laboratory.

3. At the conclusion of the field test, plants remaining at the test site were plowed into the soil. The test plot and surrounding area were watered thoroughly for 1 week to rapidly eliminate the bacterium from the soil. Because the proposal to release *X. campestris* in the environment was determined to be a major federal action, an Environmental Assessment was prepared (in accordance with the requirements of NEPA) with a Finding of No Significant Impact.

6.2   Field Test of a Vaccinia-Vector Recombinant Rabies Vaccine

On June 7, 1991, APHIS approved a request from the Wistar Institute of Anatomy and Biology to conduct a field test with a live experimental vaccinia-vector rabies vaccine (designated as V-RG) for raccoons. The field trial was conducted on state-owned land in Sullivan County, Pennsylvania. The vaccine was developed using recombinant DNA techniques to insert the rabies virus coat glycoprotein gene into the vaccinia virus genome. The foreign gene encodes the rabies virus glycoprotein responsible for induction of rabies virus neutralizing antibodies. The insertion inactivated the vaccinia virus thymidine kinase (TK) gene, which is involved in the pathogenicity of the virus.

The specific objective of the proposed field test was to collect information on the safety and efficacy of V-RG under field conditions. The vaccine was offered to raccoons by the oral route. The vehicle for vaccine delivery was a vaccine-bait unit placed in a polyethylene bag that contained a slurry to enhance bait attractiveness to raccoons and repellency to humans. The vaccine was shown to be safe and efficacious in a variety of host animal studies carried out in containment.

The following critical areas were examined in detail:

1. the molecular properties of the vaccine master seed; e.g., the cloning site, the identity of the gene located at the cloning site, and the function of the donor construct in the recombinant organism

2. the safety characteristics of the vaccine; for instance, reversion to virulence potential, genetic and phenotypic stability, effect of overdosing, tissue tropism in susceptible hosts, and potential effect of horizontal gene transfer and/or recombination with wild-type pox viruses

3. the ecological characteristics of the vaccine microorganism; e.g., the host/range specificity, shed/spread capability, identity of biological vectors, and the survivability of the vaccine microorganism in the target environment

General concerns raised by this proposal included the broad host range of vaccinia virus, its survivability in the environment, and its ability to infect humans. Each concern was properly addressed by Wistar with the following studies, mitigative procedures, and monitoring:

1. The safety of V-RG to nontarget animal species was thoroughly evaluated as demonstrated in Table 4.

2. Potential human health risks were assessed by the Vaccinia Subcommittee, National Vaccine Program, Department of Health and Human Services. Personnel working on the study were immunized against vaccinia in compliance with U.S. Public Health Service guidelines for individuals working with recombinant vaccinia viruses.[8]

3. Each vaccine-bait unit was checked for 10 days after placement. Any remaining field baits, bags, and associated debris on day 10 were removed for proper disposal.

4. After vaccine-bait removal, monitoring of the field test area was instituted for a period of 12 months. This included live-trapping of raccoons and nontarget animals. Blood samples were collected, and a limited number of animals were euphanized for serology, virus isolation, postmortem analysis, and histopathological examination.

Because the proposal to release V-RG in the environment was determined to be a major federal action, the procedural requirements of NEPA were applied. APHIS prepared an Environmental Assessment and advised the public that APHIS would hold a public meeting in the State of Pennsylvania to discuss the Environmental Assessment and Finding of No Significant Impact (56 FR 19635–19636). The facts supporting the Finding of No Significant Impact are summarized in Table 5.

TABLE 4

Summary of Safety Trials of Vaccinia-Vector Recombinant Rabies Vaccine in Nontarget Species

Species	Number	Dose (Range, pfu)*	Route	Observation Period (range, days)
Class Mammalia				
Order Marsupiala				
Family Didelphidae				
Opossum	6	10^1	oral	30
(*Didelphis virginianus*)				
Order Carnivora				
Family Canidae				
Red fox	37	10^6–10^8	oral	0.5–2.8
(*Vulpes vulpes*)	8	$10^{8.5}$	bait	90
	6	$10^{8.5}$	GI	90
	2	$10^{8.5}$	ID	90
	2	10^8	ID	28
	1	10^8	SC	28
	20	10^4–10^8	oral	28–365
	5	10^8	bait	28
	2	$10^{7.6}$	ocular	30
	2	$10^{7.6}$	intranasal	30
Fox cubs	13	$10^{7.2}$	oral	33–365
(*V. vulpes*)				
Dog	9	$10^{4.6}$–$10^{8.6}$	SC	28
(*Canis familiaris*)				
Grey fox	9	$10^{8.6}$–$10^{9.6}$	oral	28–30
(*Urocyon cinereoargenteus*)				
Coyote	10	$10^{7.9}$	bait	90
(*Canis latrans*)				
Family Felidae				
Bobcat	3	10^9	oral	30
(*Lynx rufus*)				
Cat	4	10^8	oral	115
(*Felis domesticus*)	9	10^4–10^8	SC	52
Family Mustelidae				
Skunk	8	10^9	bait	90
(*Mephitis mephitis*)	8	10^9	GI	90
	6	$10^{8.3}$	ID	90
	3	$10^{8.3}$	IM	90

TABLE 4 (CONTINUED)

Summary of Safety Trials of Vaccinia-Vector Recombinant Rabies Vaccine in Nontarget Species

Species	Number	Dose (Range, pfu)*	Route	Observation Period (range, days)
River otter (*Lutra canadensis*)	3	10^9	oral	30
Mink (*Mustela vision*)	7	$10^{7.7}$	oral and ID	180
Ferret (*Mustlea putorius*)	4	10^8–10^9	oral	28
European badger (*Meles meles*)	6	$10^{8.3}$	oral	45
Family Ursidae				
Black bear (*Ursus americanus*)	3	$10^{8.8}$	oral	30
Order Artiodactyla				
Family Bovidea				
Cattle	10	10^8	SC	120
(*Bos taurus*)	12	10^8	ID	35–120
	1	10^8	SC	35
	1	10^8	IM	35
Sheep (*Ovis ovis*)	4	10^7	oral	30
Family Suidea				
Wild boar (*Sus scrofa*)	4	$10^{8.3}$	oral	88
Family Cervidae				
White-tailed deer (*Odocoileus virginianus*)	4	10^9	oral	30
Order Lagomorpha				
Family Leporidae				
European rabbit	4	$10^{8.3}$	ID	14
(*Oroctolagus* sp.)	2	$10^{7.8}$	ID	21
	2	$10^{7.8}$	IM	21
	2	$10^{7.8}$	SC	21
	2	$10^{7.8}$	oral	21
	3	$10^{7.6}$	ID	180
Order Rodentia				
Family Muridae				
House mouse	12	$10^{8.3}$	ID	14
(*Mus musculus*)	12	$10^{7.7}$	FP	14
	12	$10^{6.9}$	oral	14

TABLE 4 (CONTINUED)

Summary of Safety Trials of Vaccinia-Vector Recombinant Rabies Vaccine in Nontarget Species

Species	Number	Dose (Range, pfu)*	Route	Observation Period (range, days)
Family Erethizonidae				
Porcupine (*Erethizon dorsatum*)	3	$10^{9.0}$	oral	30
Family Scluridae				
Groundhog (*Marmota monax*)	10	$10^{7.9}$	oral	90
Grey squirrel (*Sciurus carolinensis*)	11	$10^{7.9}$	oral	90
Family Cricetidae				
Cotton rat (*Sigmodon hispidus*)	8	10^{8}	oral	1–30
Marsh rice rat (*Orzomus palustris*)	7	10^{8}	oral	60
Syrian hamster (*Mesocricetus auratus*)	12	10^{7}	IM	30
Field vole (*Mircotus agrestis*)	1	$10^{6.5}$	oral	35
Meadow vole (*Mirotus pennsylvanicus*)	26	$10^{7.9}$–10^{9}	oral	30–90
Common vole (*Microtus arvalis*)	2	$10^{6.5}$	oral	35
Bank vole (*Clethrionomys glareolus*)	13	$10^{6.3}$	oral	35
Water vole (*Arvicola terrestris*)	5	$10^{6.5}$	oral	35
Deer mouse (*Peromyscus maniculatus*)	10	$10^{9.0}$	oral	90
European field mouse (*Apodemus* sp.)	4	$10^{6.3}$	oral	41
Yellow-necked mouse (*Apodemus flavicollis*)	7	$10^{6.5}$	oral	41
Wood mouse (*Apodemus sylvaticus*)	27	$10^{6.3-6.5}$	oral	28–43
Class Aves				
Order Falconiformes				
Family Accipitridae				
Red-tailed hawk (*Buteo jamaicensis*)	6	10^{8}	oral	30
Kestrel (*Falco tinnunculus*)	4	10^{8}	oral	30
Carrion crow (*Corvus corone*)	17	10^{8}	oral	28

TABLE 4 (CONTINUED)

Summary of Safety Trials of Vaccinia-Vector Recombinant Rabies Vaccine in Nontarget Species

Species	Number	Dose (Range, pfu)*	Route	Observation Period (range, days)
Common buzzard (*Buteo buteo*)	8	10^8	oral	30–45
Order Charadriiformes Family Laridae Ringbill gull (*Larus delawarensis*)	2	$10^{7.9 \text{ or } 8.1}$	oral	90
Order Strigiformes Family Strigidae Great horned owl (*Bulbo virginianus*)	8	10^8	oral	30
Order Passeriformes Family Grallinidae Magpie (*Pica pica*)	7	10^8	oral	28
Family Corvidae Jay (*Garrulus glandarius*)	2	10^8	oral	28

Note: Abbreviations — PFU, plaque-forming unit; SC, subcutaneous; ID, intradermal; IM, intramuscular; FP, footpad.

TABLE 5

Finding of No Significant Impact from *Federal Register* 56(82), April 29, 1991

- Genetic engineering procedures were employed to incorporate the rabies glycoprotein gene within the thymidine kinase (TK) locus of vaccinia virus. The resultant vaccinia virus cannot induce rabies.

- The V-RG virus was shown to cause no adverse clinical signs or gross or histopathological lesions, yet it was fully capable of eliciting an immune response that protected a variety of species from virulent rabies virus challenge. The V-RG virus was unable to evoke antibodies to other rabies virus structural proteins.

- The TK gene insertion is a stable characteristic of the V-RG virus with a very low probability of loss or reversion.

- The V-RG virus does not contain an oncogene or cancer-causing substance and does not contain any new genetic information to enhance the likelihood of its becoming oncogenic.

- Biological transmission of the V-RG virus could not be demonstrated in laboratory studies involving more than 40 species of domesticated and wild mammals and birds.

TABLE 5 (CONTINUED)

Finding of No Significant Impact from *Federal Register* 56(82), April 29, 1991

- In rare instances, contact (mechanical) transmission between animals was observed immediately after oral administration of the V-RG vaccine. No adverse effects from these transmissions were observed, and no advance outcomes are expected from similar exposures during the field trial.

- Contained laboratory experiments demonstrated that the V-RG vaccine is nonpathogenic, safe, and efficacious in a variety of laboratory animal model systems and a number of target and nontarget species, including the major terrestrial wildlife and domestic animal reservoirs of rabies.

- Previous field trials of the V-RG vaccine in Europe and Virginia have not demonstrated adverse effects of V-RG for workers, local human populations, or wildlife.

- In the proposed field trial, mitigations will be in place to minimize the potential impact of the V-RG vaccine and the field trial itself on humans and the environment. These measures will include immunization of project workers and restricted access to the trial site during the 2-week initial period of vaccine-bait placement.

- Monitoring of wildlife and human population in the area of the field trial will continue for 1 year after initiation of the trial.

7   EPILOGUE

Biotechnology Permits, BBEP has received far fewer applications for the release of genetically engineered microorganisms under 7 CFR 340 than for genetically engineered crop plants. The microorganisms under 7 CFR 340 which have been modified and tested fall into three categories of potential usefulness to agriculture with a large benefit to risk ratio: (1) engineered biocontrol agents, (2) engineered microbes to study biological or disease processes, and (3) nitrogen-fixing *Rhizobium*. The performance of the genetically engineered bacteria in the field has been disappointing; they have actually proven to be less effective than nonengineered alternatives. The problem observed with all of the introductions is the lack of success in the environment. An engineered characteristic in an isolated microbe introduced into a foreign environment (i.e., a different geographic location), especially soil, seems to impart a selective disadvantage. The organism can no longer compete successfully for a niche in the milieu of microorganisms in soil or host plants. In the case of biocontrol agents, the balance between aggressive pathogenesis and the ability to compete against other microbes in the nonhost environment is critical. Superimposed in this equation are changing weather conditions.

Corporations are reluctant to invest in long-term development programs to overcome these problems when efficacy is in question. For example, there have been few introductions of *Rhizobium*, despite the fact that APHIS has issued an opinion paper declaring it to be a nonregulated article. The chances of success may be increased if an organism is selected from a specific environment and then reintroduced to the same environment following genetic manipulation.

To date, all veterinary field tests have been with live genetically engineered virus vaccines. The main barrier to industry in the development of genetically engineered bacterial vaccines has been the inability by some to remove or exclude genetic markers (i.e., antibiotic resistance genes). Recent innovations in biotechnology may eliminate this barrier in the near future.[9,10]

Other impediments include the shed/spread capability of orally administered bacterial vaccines, often requiring additional data to properly identify the ecological characteristics of these microorganisms. Nonetheless, the risk analysis method developed by Veterinary Biologics, BBEP should provide a credible process for identifying and characterizing potential risks associated with the use of genetically engineered bacterial vaccines in the field as they become available.

ACKNOWLEDGMENTS

We thank Dr. Sivramiah Shantharam for his technical input. We gratefully acknowledge Laurie Foudin for critical and constructive reading of the manuscript and Linda Lightle for excellent secretarial assistance.

REFERENCES

1. Espeseth, D. A., Joseph, P. L., Shibley, G. P. and Gay, C. G., Current USDA licensing requirements for recombinant DNA-derived animal biological products, *Rev. Sci. Tech. Off. Int. Epiz.*, 7(2), 255, 1988.
2. Gay, C. G., Current USDA procedures for licensing biotechnology-derived veterinary biologicals, *Dev. Biol. Stand.*, 79, 65, 1992.
3. Animal and Plant Health Inspection Service, User's Guide for Introduction of Genetically Engineered Plants and Microorganisms, Technical Bulletin No. 1783, United States Department of Agriculture, Washington, D.C., 1991.
4. Cohrssen, J. J. and Covello, V. T., Risk Analysis: A Guide to Principles and Methods for Analyzing Health and Environmental Risks, U.S. Council on Environmental Quality, Executive Office of the President, Washington, D.C., 1989.
5. National Research Council, *Risk Assessment in the Federal Government: Managing the Process*, National Academy Press, Washington, D.C., 1983.
6. National Research Council, *Issues in Risk Assessment*, National Academy Press, Washington, D.C., 1992.
7. Joseph, P. L., Espeseth, D. A., Shibley, G. P., and Gay, C. G., Environmental considerations for field testing and licensing of live genetically-engineered veterinary biologics, *Rev. Sci. Tech. Off. Int. Epiz.*, 7(2), 285, 1988.

8. U.S. Department of Health and Human Services, Public Health Service, Vaccinia (small pox) vaccine: recommendations of the Immunization Practices Advisory Committee (ACIP), *MMWR*, 40, No. RR-14, 1991.

9. Nakayama, K., Kelly, S. M., and Curtiss, R., Construction of an ASD+ expression-cloning vector: stable maintenance and high level expression of cloned genes in a *Salmonella* vaccine strain, *Biotechnology*, 6, 693, 1988.

10. Stover, C. K., de la Cruz, V. F., Fuerst, T. R., Burlein, J. E., Benson, L. A., Bennett, L. T., Bansal, G. P., Young, J. F., Lee, M. H., Hatfull, G. F., Snapper, S. B., Barletta, R. G., Jacobs, W. R., and Bloom, B. R., New use of BCG for recombinant vaccines, *Nature*, 351, 456, 1991.

Chapter 6

ENGINEERED VIRUSES IN AGRICULTURE

Ilan Sela

TABLE OF CONTENTS

0-8493-4465-4/96/$0.00+$.50
© 1996 by CRC Press, Inc.

1   INTRODUCTION

This chapter relates to the transformation of plants with DNA sequences (viral as well as nonviral) which alter host-virus relationships, the use of transformed plant cells to study viral gene functions, and the potential use of these transgenic organisms for plant protection. This chapter is neither a compendium nor a listing of the available literature. Examples were picked for every topic, to illustrate particular subjects, and as a basis for discussion.

2   VIRUS-RELATED GENETIC ENGINEERING IN PLANTS

In general, plants are transformed with recombinant DNA sequences to study the mechanisms by which a certain sequence functions in the cell or to study interrelationships among genes. Plants are also transformed in an attempt to produce better crops, by introducing a trait which might improve a desired quality, or to engender plant resistance to a disease or other stresses. Plants have also been engineered in an attempt to exploit them as large-scale, inexpensive producers of the cloned gene product (gene farming).

In the narrower scope of virus-related engineering, mutated and chimeric viral genomic sequences are often used in the course of studying viral-gene functions and in an attempt to convert viral sequences into efficient expression vectors. When the role and mode of action of an individual viral gene (or parts thereof) are studied, that gene is often introduced into and expressed in the plant. Plant transformations by viral and nonviral sequences are also carried out to produce engineered virus-resistant plants.

This chapter concentrates on virus-resistant transgenic plants. However, observations made through the study of virus-resistant transgenes, and the mechanisms underlying this phenomenon, have contributed to the understanding of hitherto understudied, incomprehensible, basic regulatory mechanisms occurring in plants, especially at the RNA level, and other aspects such as plasmodesmal function in cell-to-cell transport.

Plants have been reported to become virus resistant when transformed with sequences related to:

- the viral capsid protein
- the viral movement protein
- viral genes for replicases (and putative replicases)
- viral genes for proteases
- satellites of viruses and defective interfering RNAs

- antisense viral sequences
- ribozymes constructed with viral sequences
- foreign nonviral, nonhost genes
- host genes

Since several good review articles on virus-resistant transgenic plants have been published in recent years,[1-10] we present only a partial account of the listed resistant transgenic plants and concentrate on a description of the general phenomenon and on examples contributing to the under-standing of mechanisms or indicating new potential approaches.

2.1 Capsid-Protein-Mediated Resistance

Capsid-protein (coat-protein)-mediated resistance in transgenic plants was first reported for tobacco mosaic virus (TMV).[11] Since then it has become an almost general phenomenon that a plant transformed by a viral coat-protein gene becomes resistant to the corresponding virus. The fol-lowing is a representative, albeit incomplete, account of viruses for which this phenomenon has been demonstrated: several tobamoviruses,[11-14] al-falfa mosaic virus (AlMV),[15] beet necrotic yellow vein virus,[16] cucumber mosaic virus (CMV),[17-19] potato virus X (PVX),[20-22] tobacco rattle virus,[23] tomato spotted wilt virus (TSWV),[24] grapevine chrome mosaic nepovirus and the related fanleaf virus,[25,26] and a number of potyviruses.[22,27-31] Protec-tion of plants transformed with the coat-protein gene of tomato yellow leaf-curl virus (TYLCV; a gemini virus) has also been reported recently.[32] This is particularly interesting since TYLCV is, to date, the only DNA virus known to relate with its transgenic host in this manner.

2.1.1 Agricultural Aspects: Field Tests, Vectors, and Selection for Further Breeding

Several field tests, as well as environmentally oriented studies, have been conducted with coat-protein-expressing transgenic plants. The over-all conclusion is quite encouraging, since most of the plants in the transgenic population maintained at least a tolerance level of protection under field conditions. But most important, crop yields from inoculated transgene fields were at the same level as those of their respective healthy (control) fields, indicating the feasibility of using transgenic resistant plants in agriculture.[13,22,33-38]

The mode of inoculation may affect the level of protection. In most cases the transgenic plants are tested for resistance by mechanical inocu-lation with the purified virus or infected crude sap, and resistant plants are selected. Inoculation by viral RNA (as discussed below) may overcome resistance, but this type of inoculation rarely occurs in nature. Vector-mediated inoculation has been tested in only a few cases. CMV-resistant transgenes were found to be protected from aphid-transmitted CMV as well.[36] Similarly, transgenic tobacco carrying the coat-protein gene of potato virus Y (PVY) were also protected from aphid-borne PVY.[39] Tobacco plants transformed by tobacco rattle virus coat protein were protected from the aphid-transmitted virus, but were susceptible to the nematode-transmitted virus.[23]

The original transgenic plants (R0) are always heterozygous. Therefore, the production of a resistant, nonsegregating population of transgenic plants requires further conventional breeding. Homozygous resistant plants can usually be obtained from the R2 generation (provided self-pollination of these plants is possible). In the case of plants transformed with the coat-protein gene, 50–100% of the R0 plants become resistant (see, for example, References 29 and 40), and a small initial R0 population is sufficient for selecting the desired phenotype. However, in some other cases (mostly of nonstructural-gene-mediated protection; see Section 2.4) only a small proportion of the transgenic plants becomes resistant. It is therefore advisable to screen a large population of R0 plants before any conclusion can be drawn as to the successful attainment of the desired phenotype.

2.1.2 Possible Mechanisms

2.1.2.1 Virion Disassembly

Via what mechanism do coat-protein-carrying transgenes engender resistance? In the case of tobamoviruses, and probably also in coat-protein-derived resistance to other viruses, a major cause is interference with an event taking place prior to the uncoating of the virus. The resistance of the transgenes can be overcome by inoculation with higher virus concentrations[11,12,14] or with the pertinent viral RNA.[13,15,41] In these cases coat protein in the transgenic plants must be expressed and present in the cells at the time of inoculation. The plant-produced coat protein may shift the equilibrium from uncoating to recoating, simply because it exceeds the amount of coat protein brought in with the invading virus. In this model, recoating cannot take place once RNA translation begins. Therefore, free viral RNA is not affected and the infection cycle continues. In the case of TMV, a striptosome system is operating; i.e., ribosomes bind to the exposed 5′ end and translation proceeds along the exposed RNA, thus displacing the coat protein.[42,43] The resistance-conferring stage occurs earlier than cotranslational stripping, since a partially stripped TMV also overcomes coat-protein-mediated resistance.[41] Further support for a mechanism involving the prevention of uncoating is provided by the elegant experiments of Osbourn et al.[44,45] A chimeric construct was made which carried the reporter gene for glucuronidase (GUS) and the TMV origin of assembly. This construct was made a pseudovirion by coating with TMV capsid protein and was used for transfection of protoplasts. GUS expression in protoplasts which had been transformed with TMV coat protein was only about 1% of that observed in nontransformed protoplasts. Transfecting the protoplasts with naked, uncoated construct resulted in a 30- to 50-fold increase in GUS activity in the transgenic protoplasts relative to the coated pseudovirions. This is a clear indication that uncoating is indeed inhibited in the transgenes, thus considerably decreasing the level of expression of the coated genome (this report also suggests that some other mechanism occurs, as referred to below). Plants expressing TMV coat protein were crossed with plants expressing RNA transcripts carrying another reporter gene (chloramphenicol transacetylase; CAT) and the TMV origin of assembly. CAT transcript coating by the endogenous coat protein was negligible. However, these plants remained protected from TMV infection. Similar observations were made when protoplasts of the two types of transgenes were fused to express both the coat protein and the above-described transcript. These data suggest that prevention of TMV disassembly, rather than recoating, is the mechanism interfering with the infection cycle of TMV. The mechanism of TMV coat-protein-mediated protection is discussed in detail by Reimann-Philip and Beachy.[46]

2.1.2.2   *Interfering with Other Coat-Protein Functions*

The inhibitory effect of the coat-protein-bearing transgene on other viruses is probably not related to disassembly alone. Inoculation of the various coat-protein-expressing transgenic plants with the respective viral RNAs does not always overcome viral resistance. This has been demonstrated for PVX,[47] potato virus S,[48] arabis mosaic virus,[49] and (in contrast to previously reported observations[15]) for AlMV.[50] As demonstrated for RNA bacteriophages,[51] coat proteins of some animal viruses[52] and a plant virus[53] may have functions other than physical protection of the genomic RNA. These functions usually act as switching signals, directing the shift from an ongoing process to the next. This is also true for AlMV. A mixture of AlMV-RNA1 and AlMV-RNA2 is infectious only when supplemented with some molecules of the viral coat protein or with AlMV-RNA4, which carries the active gene for the coat protein.[54] This indicates that coat protein is essential for a crucial step(s) during the AlMV replication cycle within the cells. Indeed, a single amino acid exchange in the AlMV coat protein was shown to abolish its ability to confer resistance in transgenic plants.[55] Specific interactions between the viral RNAs and the coat protein were demonstrated: the unmodified coat protein in transgenic tobacco did not complement the viral RNA (1+2+3), which remained noninfectious. The mutated coat protein, however, supported RNA infectivity. Both coat proteins (unmodified and mutated) bound to the RNA, as indicated by gel-shift assays. This experiment very strongly indicates that the AlMV coat protein actively participates in a viral replication step that occurs after uncoating. Other pieces of evidence suggest that the coat protein is involved in yet a later stage of viral replication, and this stage is interfered with in coat-protein-bearing transgenes. Protoplasts transformed by the TMV U1 coat protein gene were, as expected, resistant to U1 infection. They were, however, susceptible to infection by the Cc (cowpea) strain of TMV or its naked RNA. However, these protoplasts also remained unprotected against infection by heterovirions reconstructed from Cc RNA and U1 coat protein.[44] Thus, assembly-disassembly equilibrium is apparently not the only protective mechanism.

A positive correlation was, however, demonstrated in several (but not all) cases between the level of expression of the coat protein and the degree of protection in transgenic plants (tobamoviruses, PVX, AlMV, and some other viruses).[9] This indicates that the protein itself must be present for resistance. Indeed, untranslatable constructs of TMV coat protein[56] and AlMV coat protein[57] genes did not confer resistance on transgenic plants. As discussed above, coat protein may be involved in various steps of viral replication, and its premature accumulation may "confuse" the normal cascade of virus replication events. In these presumably regulatory steps, a very small amount of "untimely" coat protein may be sufficient. In cases where a correlation between coat-protein accumulation and resistance was not observed (e.g., CMV,[18] potyviruses,[22,27] TSWV[53,58]), a mechanism requiring the actual coat protein may coincide with any other mechanism, such as RNA-mediated protection (see below).

The proposed, above-discussed mechanisms are based on interactions between coat protein and a specific target (be it the viral RNA or a viral or host protein). It is therefore not surprising that coat-protein-mediated protection is of a relatively narrow range. In some cases it protects from infection by the corresponding virus and closely related strains, but not from distant strains of the same virus.[14,44,55,59] Potyvirus resistance exhibits a wider range of protection: plants may (but don't always) become resistant to a number of potyviruses, regardless of the origin of the recombinant potyvirus coat protein.[27,28,60] Only one report demonstrated wide-range, limited protection: inoculation of plants transformed by TMV U1 coat protein with CMV, PVX, PVY, and AlMV resulted in delayed symptom appearance, whereas AlMV-coat-protein-expressing plants exhibited such a delay only when challenged with CMV or PVX.[61]

2.1.2.3 RNA-Mediated Resistance

Plants transformed with untranslatable coat-protein gene sequences of potyviruses[62-67] and TSWV[68-70] exhibited resistance, even in the absence of the protein itself. This is very reminiscent of a more general phenomenon that has come to light in recent years, i.e., co-suppression. In many (but not all) instances, when a second copy of a native gene is introduced into a plant, expression of both the native and the introduced genes is suppressed. This was first observed when expression of *Agrobacterium tumefaciens* T-DNA was silenced upon transfer of multiple T-DNA copies into the plant.[71] *Petunia hybrida* white flowering mutants exhibited a range of colored phenotypes when transformed with the AI gene of maize, which promotes the production of pelargonidin. However, some of the transgenic plants remained white. The colored phenotypes were mostly transformed with a single copy of the AI gene, whereas most of the white transgenes contained multiple-copy AI.[72,73] In both cases expression inactivation was attributed to the presence of multiple copies of the respective gene, which resulted in promoter methylation. This phenomenon was also observed with flavonoid genes of petunia,[74] the nopaline synthase gene,[75] and several others (as reviewed and discussed by Kooter and Mol[76] and by Finnegan and McElroy[77]). Besides methylation, various other mechanisms have been suggested for the phenomenon of co-suppression, including incidental transcription of the complementary DNA strand of the inserted gene, providing down-regulating antisense RNA.[78]

With plant RNA viruses, however, co-suppression due to DNA:DNA interactions is irrelevant, unless the interaction in the virus-infected transgenic plant is mediated in *trans* by the viral RNA. RNA-mediated gene silencing has therefore been postulated. Most of the relevant data come from studies with tobacco etch virus (TEV).[66,67] TEV-coat-protein-transformed plants may not initially become resistant. Resistance, however, was induced by TEV infection and developed gradually. When the transformed plant reached an antiviral state, the steady-state level of the endogenous mRNA, as well as that of the viral RNA, was reduced to 5–10% of that of control plants. Transcription was not affected at all, and the steady-state level of the transgene primary transcripts remained unaltered. The silencing of messenger RNA (mRNA) is therefore a cytoplasmic event. This mechanism might involve the production of antisense RNA (in contrast to the above-mentioned nuclear antisense mechanism) in the cytoplasm by an RNA-dependent RNA polymerase. Short double-stranded RNA (dsRNA) segments target the RNA for degradation (see Section 2.6). If sense post-transcriptional silencing is a general phenomenon, it implies the occurrence of RNA-dependent RNA polymerase in plants. Promoter for the synthesis of a complementary RNA strand is likely to reside near the 3´ end of the template strand. The same replicative mechanism is believed to be involved in both orientations of the viral RNA. Since the 5´ end of the positive strand determines the 3´ end of the negative one, both ends of the viral strand should be important for viral RNA replication.

Therefore, the observation that transgenic tobacco transcribing short antisense sequences to the 5′ end of TMV RNA[79] or the 5′ sense sequences of PVY RNA[80] are resistant may also be attributed to a similar mechanism, i.e., the signaling of partially dsRNA for degradation. This type of sense suppression does not require the gene protein product itself, and truncated or untranslatable RNAs may bring about resistance. This should, however, be a very specific mechanism, and indeed the broader resistance engendered by coat protein in the case of potyviruses was not observed in this case: the plant became protected only from a cognate virus.

2.1.2.4   Effect on Virus Movement

TMV coat protein may have an effect on the long-distance movement of the virus within the plant, since a virus with a mutated coat protein conferred a hypersensitive reaction in *Nicotiana sylvestris* to a TMV strain that otherwise would have spread systemically.[81] A gene for gene interaction must be involved between the viral coat protein gene and a host gene, since coat protein mutations did not overcome the localizing effect conferred by the tobacco N gene (a mutation at the 126-kDa TMV gene was implicated for N-gene-mediated localization[82]). In any event, in a particular genetic environment, the coat protein may influence viral movement. Such an effect has indeed been indicated: when transgenic plants were grafted onto unmodified plants and the rootstocks were inoculated with TMV, the movement of the virus into the graft was reduced as compared to the control. Furthermore, when a section of a coat-protein-bearing transgenic plant was grafted between two sections of unmodified tobacco, viral spread from the lower part to the apical (unmodified) part was also reduced, provided the intermediate transgenic graft carried a leaf.[83] A similar phenomenon has also been reported for potato virus S.[48]

2.1.3   A Comment about Stability

In several cases of natural or engineered resistance, a single amino acid substitution in the viral coat protein was sufficient to make or break resistance. This was reported for TMV (with regard to its interrelationship with the N′ gene of *N. sylvestris*),[81] for PVX (with regard to its interrelationship with the Rx gene of potato),[84] and for the engineered resistance of AlMV[55] and CMV.[85] This should raise serious concerns as to the genetic stability of the resistant phenotypes, especially under field conditions.

2.1.4   Conclusion

Coat-protein-mediated resistance appears to be a general phenomenon. When tested, it was also effective under field conditions, protecting the plants from aphid- and whitefly-borne virus inoculation as well.[32]

Other vectors (such as trips) should be tested, especially when considering TSWV resistance. One report implies that transgenic plants are not protected from nematode transmission.[23] The protection range is quite narrow and is at best effective among members of the same viral group. The delay found in interviral combination does not seem to be of practical use. Coat-protein-mediated resistance stems from interference at various levels of the viral cycle within the cells, and a multiple-mechanism scheme is implied. In some cases, the major inhibited step occurs very early on, intervening in uncoating by the actual endogenous transgene protein, but when this step is bypassed for some reason, the other mechanisms may still maintain the plant's resistance. In some cases, a sense suppression mechanism at the RNA level may be involved and the actual presence of the endogenous coat protein is not required.

2.2 Movement-Protein-Mediated Resistance

Many plant viruses code for a protein which is involved in facilitating viral movement from cell to cell within the infected plant. Many viruses carry the gene for such a movement protein, including TMV,[86,87] cowpea mosaic virus,[88] brome mosaic virus (BMV),[89] red clover necrotic mosaic virus,[90] AlMV,[91] tobacco rattle virus,[92] and cauliflower mosaic virus (CaMV, a dsDNA virus; the movement protein, however, is believed to act on its RNA transcripts).[93]

A number of review articles describe the various aspects of viral movement proteins and viral transport across cell membranes.[94-99] The 30-kDa gene product of TMV was the first to be directly identified in viral movement and is the best studied. Several TMV temperature-sensitive mutants were found to be unable to spread at nonpermissive temperatures. However, co-infection with a wild-type virus enabled movement of the mutant, thus demonstrating complementation.[100] All transport-defective TMV strains were mutated at the 30-kDa-protein gene,[87,101,102] and, reciprocally, a wild-type TMV was converted to a transport-defective one by *in vitro* mutagenesis of the 30-kDa gene.[87] Finally, movement of a transport-defective mutant was complemented and restored in transgenic plants expressing the nonmutated 30-kDa protein.[86,103]

Plasmodesmata, the natural plasma bridges between adjacent plant cells, have been shown to be involved in viral transport. The 30-kDa protein has been seen to accumulate at or near plasmodesmata of infected and 30-kDa-expressing transgenic plants, especially in their cell wall fraction.[104-107] The 30-kDa protein directly or indirectly changes the exclusion properties of the plasmodesmata. Normally plasmodesmata only allow the transport of low-molecular-weight compounds (less than 1000 Da, 1.5–2.0 nm in diameter).[105-108] However, the size of TMV particles

(18 × 300 nm) and of free-folded TMV RNA (10 nm in diameter, as cited by Citovsky and Zambryski[99]) indicates that the movement protein must increase plasmodesmal permeability to allow passage of these forms. Indeed, plasmodesmata in transgenic leaves were found to be permeable to indicator dyes 5–6 nm in diameter.[105] Microinjection of a recombinant 30-kDa protein into cells engendered a permeability limit of up to 9 nm.[109] Even at the highest permeability, the modified plasmodesmata are still incompatible for transfer of TMV or TMV RNA. Therefore, the movement protein must affect viral transport in yet another way. Indeed, several movement proteins have been characterized as single-strand nucleic-acid-binding proteins.[90,109-112] In contrast to other known single-strand binding proteins, the 30-kDa protein also binds to RNA and, like the others, is sequence nonspecific. A model in which the TMV movement protein interacts with the naked TMV RNA to make a long and thin (2 nm) transport complex was suggested accordingly.[112] The movement protein carries two sets of domains: one for binding to plasmodesmata (or to a cellular protein that is then activated to bind to plasmodesmata), increasing the latter's exclusion limit, and another which binds to the viral RNA and shapes it for transport through the "open" plasmodesmata. Since RNA binding is sequence nonspecific, a certain movement protein may complement a defective one, even of a quite distant virus (for example, BMV movement protein induces resistance to TMV in transgenic plants).[113]

The movement protein of cowpea mosaic virus (a comovirus) interacts with the plant quite differently. The cells develop tubular structures, packed with virions, that span across the plasmodesmata into neighboring cells,[114,115] and the intact virions are thus transported.[88] A tobamovirus movement protein can complement comovirus transport, as has been shown for red clover mottle virus (RCMV).[116] This was not the case in transgenic plants: the 30-kDa protein of the transgene did not complement the movement of RCMV. However, mixed infection of RCMV and a temperature-sensitive TMV strain defective in movement protein did result in complementation, even at the restrictive temperature.[117] The 30-kDa protein has been suggested to prime tobacco to allow movement of RCMV, but another viral product is also necessary to execute transport. However, it is certain that TMV- and comovirus-type modes of plant-movement protein interaction, seemingly radically different, share common characteristics and may partially complement each other.

Since movement protein interacts with both plasmodesmata and viral RNA, it is very likely that a modified movement protein which is unable to support transport would still be capable of interfering with the unmodified-movement-protein-mediated transport of a virus, thus engendering resistance. This has been demonstrated in two cases[113,118] where protection was also demonstrated against distant viruses of the same group (tobamoviruses). Since the modified movement protein interacted (albeit differently) with the transgenes' plasmodesmata, it is expected to confer resistance to a wide spectrum of viruses. However, this aspect has not yet been dealt with. The prospect of achieving a multivirus-resistant transgenic plants is most promising.

2.3   Resistance in Plants Transformd by Sequences of Viral Replicase-Associated Genes

In recent years quite a large body of evidence has been accumulated indicating that plants which carry sequences of viral genes encoding proteins of replicase-associated functions become resistant. It should be noted early on that the majority of these gene constructs do not produce the respective functional protein, and until further data are accumulated, functional-replicase-mediated protection is the exception rather than the rule. Plant viral gene products are considered to be part of a larger replicating complex, also including host constituents (for example, see Young et al.[119]), and the term "replicase" is, therefore, somewhat misleading. The virus-encoded proteins may actively participate in replication, or they may have a regulatory role or a targeting role. Hence, replication-associated proteins would seem to be a better term, although they are nevertheless referred to as replicases. Several reviews on replicase-mediated resistance have been recently published.[120-122]

Replicase-related resistance has been reported for TMV,[123,124] CMV,[125] PVX,[126,127] PVY,[128] pea early browning virus,[129] and cymbidium ringspot virus.[130] In at least two cases, however, plants transformed with a full-length viral replicase gene (AlMV[131] and BMV[132]) did not exhibit resistance. Replicase-mediated resistance was first reported in 1990,[123] but the data accumulated to date are insufficient to draw general conclusions. It seems, however, that this type of resistance is stronger than that mediated by coat protein, as it is maintained when higher titers of viral inoculum are applied.[123-125,127] In most cases tested, the engendered resistance was quite specific to the virus of origin and did not extend to other viruses of the same group, even when closely related,[123,129] except for one case in which a broad spectrum of resistance to tobamoviruses was demonstrated.[124]

In several cases the resistance-engendering replicase gene was defective: a considerable (bacterial) insertion was accidentally placed in a TMV replicase construct,[124] a deletion was made that allowed expression of only 75% of the CMV gene product,[125] or a dysfunctional mutant of PVX replicase gene was used for transformation.[127] Resistance conferred by constructs which were unable to produce a functional protein leads, at this time, to the conclusion that RNA-mediated protection (see Section 2.1.2.3) is involved in these cases. It is more difficult to assess the situation in the case of the read-through proteins of the TMV[123] and pea early browning[129] replicase systems. The TMV replicase sequence used to confer resistance included an open reading frame (ORF) for a putative protein (54 kDa) that has never been found in infected plants, not even in plants transformed with a seemingly translatable construct of this gene. Known TMV-encoded components of the replication system of this virus are the 126-kDa and the 183-kDa proteins, the latter being a read-through product of the former. Neither of these unmodified or nontruncated genes conferred resistance in transgenic plants.[123,124] Because the putative 54-kDa protein should result from an ORF within the read-through portion of the 183 kDa protein, it can be considered a truncated 183 kDa protein and can be listed among the dysfunctional, resistance-engendering replicase sequences. On the other hand, the 54-kDa protein might be produced at subdetectable levels, in which case it probably plays a regulatory role. An indirect indication that a 54-kDa protein may be expressed *in vivo* is the presence, in plants, of the respective subgenomic RNA and dsRNA replication forms.[133,134] In an attempt to assess the requirement of the translated 54-kDa protein for viral resistance, transient expression in protoplasts was employed.[135] The 54-kDa protein was not detected in infected protoplasts. However, the unmodified or slightly modified "gene" was capable of inducing resistance, whereas a frameshift mutation, directing synthesis of only part of the protein, did not engender resistance. The authors implied that the 54-kDa protein, and not RNA, is responsible for resistance. The frameshift mutant was capable of translating only 20% of the protein. Therefore, although protein-mediated resistance may be implied, it need not necessarily be functional. In addition, a protein which is one fifth of the size of the original may not be at all relevant to a physiological phenomenon and may have no bearing on the observed resistance.

Two cases of resistance mediated by full-length replicase sequences (not necessarily functional)[126,130] and two cases in which full-length transformation did not confer resistance[131,132] have been reported. The question of whether a dysfunctional protein is required therefore remains open. At present, RNA-mediated resistance is more feasible. If applicable, it would indicate (due to the narrow spectrum of resistance) that very specific short sequences are implicated in sense suppression. As in coat-protein-mediated protection, the phenomenon is probably multifarious, and even if a general type of mechanism is involved, individual cases may digress from it.

2.4   Resistance Engendered by Viral Proteases

Three papers reporting protease-mediated resistance, all dealing with potyviruses, have been published recently.[136-138] The potyvirus genome (as exemplified by PVY) consists of a 9.7-kilobase (kb) RNA in a sense orientation, which is translated into a single 350-kDa polyprotein.[139] The primary polyprotein is self-processed to give seven mature viral proteins[140,141] (the N-terminal portion of the NIa protein is the genome-linked protein, VPg, attached to the 5′ end of the viral RNA; hence it can be considered as an eighth protein). The order (5′ to 3′) of the PVY cistrons is: P1, helper component (HC), P3, cylindrical inclusion (CI), NIa, NIb, and coat protein. NIa is the major potyvirus protease, cleaving the nascent primary protein into several mature proteins.[142,143] Proteins directed by the first two cistrons, P1 and HC, have also been shown to be proteases, participating in self-processing of the potyviral polyprotein. The P1 product cleaves between the first and second proteins,[144] and the HC (required for insect transmission of potyviruses) is also a protease, cleaving at its own C terminus.[145] The processing of the proteins derived from the 5′ end is independent of NIa and does not require the rest of the potyviral genome. Thus, a construct carrying only the first three cistrons of PVY was expressed and correctly processed both in bacteria[146] and in plants.[138] In the course of studying PVY processing, a construct carrying the last three cistrons of PVY (NIa–NIb–coat protein) was introduced into plants. Western blot analyses with antibodies to the coat protein demonstrated that this construct is self-processed in the transformed plant. The resultant transgenes were tested for resistance (which was expected due to the presence of the coat protein cistron). The obtained resistance was, however, stronger than expected: a high virus inoculum (30–100 g/ml) did not overcome resistance, and the plants remained resistant even following exposure to 35°C.[136,147] Consequently, plants were transformed with each cistron separately. A small proportion of the NIa-transformed plants (3 lines out of 50) were found to be fully protected from PVY, even 60 days after inoculation, and two additional lines were partially protected, exhibiting a delay in symptom appearance.[136] Maiti et al.[137] analyzed whether the expression of other nonstructural potyviral genes (of tobacco vein mottling virus) would confer resistance to that virus. NIa indeed engendered resistance, whereas CI did not. The NIa-derived resistance was strong (did not break down when inoculated with 100 μg/ml of virus) and, in contrast to coat-protein-mediated resistance for potyviruses, was specific only to the virus from which the gene had been derived.

As part of the study on the independent self-processing of the N-terminal portion of the PVY polyprotein, the first three PVY cistrons were introduced into plants. Two transgenic lines (out of 50) were found to be resistant to PVY.[138] This construct carried two genes coding for proteases. At this point, however, it is premature to conclude that the proteolytic nature of two of the three genes has any direct bearing on protection. The dual nature of the HC (insect transmissibility agent and protease) is of interest, especially as HC is known to interact with coat protein.[148]

In a reciprocal experiment,[149] complementary DNA (cDNA) fragments representing the entire PVY genome were introduced into plants, constructing, in fact, a genomic library of the virus. The transgene population was inoculated with PVY, and resistant plants were selected. Analysis of the inserts of these resistant plants mapped resistance to the NIa region and to the 5′ sequences (including the 5′ untranslated region [UTR]) of PVY. Thus, the reported protease and 5′-derived resistance was reciprocally corroborated. In another study, plants transformed with a truncated P1 (protease) cistron of PVY developed immunity to low concentrations of PVY and tolerance to high viral concentrations.[150]

In the three reported cases of protease-mediated potyvirus resistance, the expressed proteases were proteolytically active.[136-138,147] A mechanism by which the plant becomes protected due to co-translational, immature processing of the primary protein of the invading virus by the endogenous protease is, therefore, possible. This is somewhat corroborated by the finding that a truncated P1 protease is less protective than the intact one.[150] However, only 5–10% of the protease-expressing transgenic plants were found to be resistant.[136,138] It is therefore impossible to assign such proteolytic activity as a general mechanism for protection. It can, however, be postulated that incidental modifications of the genomic construct bring about the expression of modified proteases which in turn engender resistance (by outcompeting the active viral proteases).

The role of VPg (which is part of the NIa cistron) in protection was also evaluated. In this respect, the NIa papers[136,137] balance each other. In the case of tobacco vein mottle virus, the transforming construct carried VPg sequences, and the possibility that the presence of VPg outside of its usual context interferes with postulated VPg functions (initiation of RNA replication, packaging) was discussed. In the case of PVY, most of the VPg sequence was deleted, and resistance due to lack of VPg was discussed. Presently, the two papers complement each other in this respect, and VPg is apparently not involved in protection.

RNA-mediated protection could still be a cause of resistance with respect to protease expression, although the only indication in favor of this is the low incidence of actually resistant transgenic plants.

2.5   Satellite-Mediated and Defective Interfering RNA-Mediated Protection

"Satellites" are virus-dependent replicating entities sometimes found in infected cells. They are unable to replicate on their own and are therefore considered to be parasites, requiring viral functions for their own replication. They show no sequence homology to their supporting virus and have not evolved from that virus. They may code for a coat protein and become encapsidated (satellite viruses) or be encapsidated by the coat protein of the supporting virus (satellite RNAs). Satellite RNAs may or may not carry ORFs (for nonstructural proteins). A number of review articles describe the general area of satellite structure and biology.[151-153]

In some instances, satellite RNA replicates via a virus-encoded replicase,[154-156] while in other cases a postulated product of its own ORF is also required.[157,158] The mechanism of satellite RNA replication has not yet been elucidated, and it is doubtful whether a single, general mechanism applies to all cases (as some satellites do not contribute a protein for replication, while others require expression of their own coded protein). However, several observations indicate the importance of 5′ satellite sequences[159] and possibly 3′ sequences[160] (apparently required for the initiation of (−) strand and (+) strand replication, respectively). Circles[161] and hammerhead structures[162,163] have also been found in some cases, implicating a rolling circle, self-cleavage mechanism, as described below for ribozymes. The satellite RNA is probably recognized by a helper virus factor (proven in some cases to be the viral replicases). The recognition site(s) is probably minimal, since satellite RNAs can be systematically mutated with infrequent loss of biological activity. This results in heterogenesis of satellite RNA upon passage in host plants[164-166] and may explain the different naturally occurring phenotypes of many of them.

Satellite RNAs may affect the pathogenicity of the helper virus (as reviewed by Collmer and Howell[167]). In many instances, satellites attenuate symptom severity (and damage), and this phenomenon is associated with reduced rates of viral replication. This has generally been attributed to competition for a component of the replication machinery between virus and satellite.[153,168] However, the satellite modulates symptom appearance and severity in other ways as well. Chlorosis and bright yellow mosaic in CMV-infected tobacco are associated with certain types of satellite RNAs. Viral accumulation in satellite-carrying plants is not different from that in satellite-devoid plants, yet symptoms are altered. The pertinent satellites have been suggested to influence chlorophyll accumulation by mobilizing the CMV coat protein into chloroplasts or by affecting a cytoplasmic component essential for chlorophyll biosynthesis, without damaging the chloroplast itself.[153,169] Satellite sequences responsible for chlorosis induction were mapped on the respective satellite RNAs at several bases around position 50.[170,171]

Some symptom-exacerbating satellite RNAs have also been found. Attention was focused on satellite-derived symptom (and damage) intensification following an epidemic in which CMV caused lethal necrosis in tomatoes. It was later found that the necrotizing entity was, in fact, a certain type of satellite RNA.[172,173] Necrogenic satellite RNAs share a highly conserved region at their 3′ end,[174,175] and necrosis modulation was mapped to the 3′ end of the satellite RNA, around position 320.[170] Some satellites carry domains for both chlorosis (mostly in tobacco and other *Nicotiana* species) and necrosis (in tomato).[170,176]

Symptom modulation depends on the interaction between a pertinent domain-carrying satellite, the virus strain, and the host plant. Several CMV-associated necrogenic satellite RNAs require the presence of RNA 2 from a strain of CMV belonging to subgroup II only.[177] Induction of chlorosis was also associated with RNA 2 of CMV of subgroup II.[178] In the case of turnip crinkle virus, the influence of the helper virus on satellite-derived symptom intensification was finely mapped.[179] A helper virus exhibiting strong amplification of symptom severity upon inoculation with a satellite RNA differed by only five bases from a viral strain which produced milder symptoms under these conditions. This difference caused a single amino acid change in the replicase gene (in the helicase domain). Satellite-modulated chlorosis of CMV in *Nicotiana* species was also found to be regulated by a host gene.[180]

Satellites, being mostly attenuators of viral infection due to competition for replicase, were among the first constructs to be introduced into transgenic plants for resistance.[181,182] Satellite-carrying transgenes should become the optimal bearers of resistance, since satellite replication is virus-induced, and their level should considerably increase upon infection. The orientation of insertion of a satellite cDNA is of no importance, since both the (–) strand and the (+) strand serve as templates for replication.[183] Several constructs of benign CMV satellites were co-inoculated with virus or introduced into plants, conferring resistance on these transgenes.[184-190] However, plants transformed with necrogenic cDNA sequences appeared normal until inoculated with CMV, upon which they became lethally necrotic rather than protected.[191,192]

Defective interfering (DI) RNAs (and DI virus particles) are quite common in animal viruses.[193] In contrast to satellites, DI RNAs are derived from their respective viral genomes by deletions or rearrangements (RNA recombination). Those modified RNAs, which are still recognized by the replication machinery of the virus, are replicated along with viral RNA. Although DI RNAs cannot exert viral functions, they compete for components of replication and thus interfere with virus infectivity.

DI RNAs occur naturally in plant cells infected with cymbidium ringspot virus,[194] TSWV,[195] and turnip crinkle virus.[196] All DI RNAs of TSWV carry an ORF (the reason for which has not yet been elucidated). They are suggested to have arisen from jumps of the viral RNA polymerase during transcription of the L segment of TSWV.[195] As in the case of satellite RNAs, transgenic plants which express DI RNAs may attenuate viral infection. This has been found to be true in plants transformed with DI DNA derived from cymbidium ringspot virus and TSWV.[194,197]

Once the structural features of the viral promoter for its respective RNA polymerase are determined, it should be possible to deliberately construct an artificial DI RNA. This possibility has been explored in depth in the case of BMV.[198,199] Artificial TMV-RNA replicons have also been produced.[200] Such artificial DI RNAs may interfere with infection.[201,202] It is therefore expected that DI-RNA-mediated protection will soon be extended to artificial DI constructs.

2.6 Resistance Conferred by Antisense RNA

In several instances, viral sequences have been introduced into transgenic plants such that the negative viral strand (antisense) is transcribed. In light of earlier information about down-regulation of gene expression by antisense RNA in cells (including plant cells), the aim was to obtain virus-resistant plants.

Antisense regulation of gene expression is a natural mechanism in prokaryotes which has also been exploited in eukaryotic cells. The general topic has been reviewed elsewhere.[203] Briefly, specific expression of a strand which is negative to a certain gene transcript may intervene in the expression of that gene at different levels. In the nucleus, antisense RNA may bind to the loop initiated by RNA polymerase and thus interfere with transcription initiation. It may also bind to the nascent transcript, preventing its elongation or proper termination. Antisense:transcript complexes may alter splicing or transport to the cytoplasm. In addition, excess antisense transcripts in the nucleus may trap the *trans*-acting proteins required for the transcription of that particular gene, thus reducing the level of its expression.[204] In the cytoplasm, hybridization of antisense RNA to mRNA may lead to translation arrest by preventing ribosomes from either binding to, or sliding along, the mRNA.[205-207] Hybridization to a certain RNA may also target this dsRNA for degradation,[208] and expression may decline due to a lower steady-state level of the respective mRNA. Antisense regulation of gene expression has been successfully exploited in plants. A few examples are the reduction of the steady-state level of transcripts to nopaline synthase,[209] inhibition of flower pigmentation by antisense for chalcone synthase,[210] and intervention in the expression of ribulose biphosphate carboxylase[211] and of tomato polygalacturonase.[212] In some cases a decrease in the steady-state level of mRNA was observed,[213] sometimes without reduction in the amount of the relevant protein[214] (indicating that only a small portion of the mRNA is required for translation of normal levels of that protein). In some other instances, the level of mRNA remained unaltered, yet the amount of protein decreased,[215,216] indicating that the translatability of the mRNA had been affected.

Most plant viruses replicate in the cytoplasm and do not go through any nuclear phase. Hence antisense viral sequences most likely exert resistance only through cytoplasmic events: either by intervening in translation or by promoting mRNA degradation. Competition for components of the viral replication machinery is also possible, since in many cases the same "replicase" transcribes the sense strand as well as the complementary one. Antisense-mediated protection has been conferred in transgenic plants against CMV,[217] PVX,[20] TMV,[78,218] and potyviruses.[67,219] In all these cases, antisense-mediated resistance was less frequent and milder than that mediated by coat protein. Strong antisense-derived protection was, however, reported for potato leafroll virus.[220] Expression of the AL1 gene (absolutely essential for viral DNA replication) of the gemini virus tomato golden mosaic virus (TGMV) in an antisense orientation engendered resistance in the transgenic plants.[221]

In cases where antisense RNA engendered resistance in transgenic plants, the antisense-affected regions were diverse: in TMV the RNA should complement the last 117 bases of TMV RNA,[218] and blocking of the replicase binding site (at the 3' UTR) was suggested. Similar reasoning applies for the 5' end of the viral RNA,[79] since it corresponds (as discussed above) to the negative-strand 3' UTR promoter. In contrast to RNA viruses, the DNA gemini viruses replicate in the nucleus. Antisense inhibition of the gemini virus TGMV could therefore exert its influence on nuclear mechanisms as well. It may also potentially confer resistance to a wider spectrum of gemini viruses, since sequences of the AL1 gene are conserved among this viral group. Indeed, the antisense expression of AL1 of TGMV also conferred resistance to beet curly top virus, which is 64% homologous to TGMV and which carries a stretch of 280 bases which is over 80% homologous to that of TGMV. The AL1 of African cassava mosaic virus is also 63% homologous to that of TGMV, but the homologous sequences are more dispersed, and no protection by TGMV-AL1 was observed.[222]

In summary, to date, antisense inhibition of viral replication has been found to be milder, and less frequent, than that of coat protein- or functional-gene-mediated protection. It also depends on the genomic sequences for which the antisense RNA is complementary. More efficient protective antisense-based systems are likely to be developed following a continuing study of viral genomic sites which are better affected by antisense.

2.7   The Potential of Ribozymes to Protect Plants from Viruses

Ribozymes derive from natural mechanisms in which RNA is self-processed, catalyzing its own specific cleavage. Whereas in all hitherto known natural cases self-cleavage was due to intramolecular folding, it is possible (as described below) to separate the catalytic core from the specific cleavage site by placing them on two different molecules, thus creating intermolecular activity resembling that of enzyme and substrate. The catalytically active RNA molecule is generally termed a "ribozyme". Since in these artificial systems the ribozyme is attracted to the proper cleavage site by base pairing, it is in fact a targeted case of antisense-mediated activity.

Precursors of ribosomal RNA were found to self-process to mature forms in *Tetrahymena*.[223] Later, several viroids and virusoids were also found to be self-processed, cleaving a unit-size RNA from the multiunit product of the rolling circle.[224] Many, but not all, of the pertinent RNAs folded into a "hammerhead" structure consisting of three stable stems and a flexible loop.[225,226] The structure carries 13 conserved bases, some of which are dispensible.[227] Two RNA molecules, carrying the required sequences, may act in *trans* and combine to form a hammerhead structure, in which case one of them (the substrate) is cleaved at a specific site. The other RNA (the ribozyme) can then bind to another such substrate and promote its cleavage.[228] The substrate RNA has only one sequence requirement other than the ability to base pair with the ribozyme strand, which is a GUX (X = C, A, or U) motif.[229] The ribozyme carries all the structural features that enable catalytic cleavage upon its binding to the substrate. Hence, by adding short complementary "arms" to the ribozyme core, it could target the desired RNA substrate provided the latter carries the GUX motif.[229]

To date, the only report in which a predesigned ribozyme has been found to be effective in protecting cells from a virus was in the case of human immunodeficiency virus (HIV).[230] Several attempts to introduce ribozymes into transgenic plants have not brought about protection, even though in some cases proper cleavage did take place *in vitro* (see, for example, Park and Hwang[231]). In the literature, successful resistance is usually referred to as "personal communication",[6,232] but not a single research paper has yet been published explicitly demonstrating ribozyme-mediated resistance in transgenic plants. Similarly, artificial ribozymes designed to cleave a reporter gene (GUS) RNA were active *in vitro*, but failed to cleave in transformed cells.[233] An indication that the ribozyme system is not yet well characterized is the observation of its extremely high specificity. Glutamine synthetase is encoded by four closely related genes in *Phaseolus vulgaris*. Hammerhead structures were constructed and targeted to cleave at a site where all four mRNAs share 18 out of 22 bases. The ribozymes, however, discriminated between the highly homologous RNAs.[234] Also, in *in vitro* experiments with potato leafroll virus RNA, ribozymes were effective at cleaving the target RNA when the substrate motif was GUC, but not when it was altered to GUA.[235] Hence the generality of the GUX motif is also questionable. If the nature of such a high degree of specificity is understood, efforts to reduce it may lead to *in vivo* protected plants as well.

2.8   Resistance Due to Nonviral, Nonhost Proteins

2.8.1   Antibody-Mediated Resistance

The concurrent development of hybridoma-derived monoclonal antibodies and gene-cloning techniques has enabled the production of recombinant antibodies and parts thereof in various expression systems.[236] Expression and assembly of functional antibodies in transgenic tobacco were first demonstrated for a murine antibody.[237] Only a few immunoglobulins were hitherto reported to be expressed in plants.[238-240] The intriguing possibility of transforming plants with antiviral antibodies has thus arisen. However, the field of plant-produced antibodies (for which Wilson[6] coined the term "plantibodies") is fairly new, and antibody-mediated protection from only two plant viruses (artichoke mottle crinkle virus[241] and TMV[242]) has been reported to date.

Antiviral antibodies appear to act via the neutralization of viral surface proteins in a way that intervenes with the initial establishment of infection. Many technical parameters for improving this type of antibody-mediated protection have yet to be determined: e.g., would increasing the level of antibody expression help? What is the best cell compartment to which the antibody should be targeted? Is a partial antibody or a single-chain transformation sufficient? Intervention in viral replication by endogenous antibodies should be further explored in yet another respect: against which viral epitope should the protecting antibody be directed? A conserved capsid-protein motif may lead to broader resistance to many strains of the same virus, or even to other, different viruses of the same group. It may also interfere with insect transmission, as evidence is accumulating for the coat protein's role in this respect (see Section 2.1.2.4). Functional viral proteins accumulate in infected cells to lower levels than structural proteins.

Therefore, antibody to a functional viral gene product might be more efficient than an anti-structural protein antibody. In addition, a broader spectrum of resistance may potentially be achieved if the antibody is directed to a conserved region of a functional protein which is shared by viruses of many groups. Replicase (as suggested by Baulcombe[10]) is a good example, since several motifs (nucleotide binding, helicase) are quite conserved. The technical parameters for such a type of protection could differ from those of capsid-targeted endogenous antibodies.

2.8.2   Resistance Due to Interferon-Related Genes

During a study of putative plant mechanisms of defense against viral infection, the stimulation of an antiviral substance(s) in virus-infected plants was demonstrated. The antivirally active fraction was designated "antiviral factor" (AVF).[243] The AVF was fractionated as a broad peak by ion-exchange chromatography,[244] but only with the advancement of high-performance liquid chromatography (HPLC) separation for proteins was AVF resolved to several different proteins.[245] A major AVF protein was purified to homogeneity and characterized as a 22-kDa glycoprotein. AVFs were very elusive and, until sensitive and accurate separation methods were developed, could only be followed by monitoring their biological activity. AVF was particularly stimulated in host plants that confined viral infection to local lesions, particularly the N-gene-carrying tobacco varieties,[246,247] thus resembling a number of the "pathogenesis-related" (PR) proteins (see Section 2.9 below). Some biological characteristics of AVF were reminiscent of animal interferons (for background data please see References 248–250).

The observation that human interferon is (transiently) antivirally active in plants as well[251] was quite accidental: interferon was used as an internal control in AVF deglycosylation studies, and surprisingly, was found to suppress TMV as well. The protective activity of interferon in plants was disputed for several years. Some groups were able to support it,[252-255] while other groups were unable to repeat the experiment.[256-258] Apparently, these inconsistencies were due to the applied interferon concentration: only low concentrations are active in plants.[259] In addition, two plant glycoproteins with antiviral activity, gp22 (22-kDa) and gp35 (35-kDa), were isolated from N-gene-carrying, TMV-infected tobacco by immunoaffinity chromatography using antibodies to human β-interferon.[260] Partial amino acid sequencing identified gp35 as one of the PR proteins: 1,3-glucanase.[261] Gp22 seems to belong to another family of PR proteins typified by osmotin (to which PR-5 also belongs), but is not identical to any of the hitherto known proteins.[262]

In animal tissues interferon stimulates several cellular pathways, including that of the dsRNA-dependent 2'-5' oligoadenylate synthetase. This enzyme polymerizes ATP to a series of oligoadenylates with an unusual phosphodiester link (2'-5' rather than 3'-5'), cumulatively designated 2'-5'A;[263] 2'-5'A activates a ribonuclease which plays a role in engendering an "antiviral state" in animal cells. Hence, 2'-5'A is in and of itself antivirally active. Also, 2'-5'A and related derivatives were found to be antivirally active in plants.[264-267] Furthermore, viral infection in plants elicited an enzyme analogous to the human 2'-5'A synthetase, as evidenced by its reaction with antibodies to human 2'-5'A enzyme and its polymerization of ATP to a short oligonucleotide, similar but not identical to 2'-5'A.[268] Interferon and 2'-5'A were reported to have some other physiological activities in plants as well.[269] The occurrence of an interferon-analogous plant system has therefore been implicated.

Several groups constructed transgenic plants expressing human interferons.[270,272] If these plants had become virus resistant they would have been protected from a wide range of viruses, due to the nonspecificity of interferon. However, the results to date are inconclusive: some plants have become protected, but most of them have not. The protected plants segregate, and resistance seems to be unstable. Plants that express interferon poorly seem to be better protected than those that express higher levels of interferon.[271,273] The physiological effects of constitutive expression of interferon are probably quite intricate, and no conclusions can be drawn at this time.

On the other hand, transgenic potato plants expressing rat oligoadenylate synthetase were shown to become PVX resistant.[274] This paper implied (with no supporting data as yet) that this is, as expected, a general type of resistance, affording protection from other viruses as well. Manipulation of the interferon system for plant protection is an intriguing aspect for further study, especially as some interferon-analogous systems may occur in plants[260,268,275,276] and may even be connected to some PR proteins.[261,269]

2.9 Potential Manipulation of Host Genes for Plant Protection

Conservative plant breeding has always relied on the native gene pool, from which genes conferring some protection (tolerance to resistance) are introduced into a commercial crop from a resistant cultivar, a related species, or a wild plant which can be sexually crossed with that crop. In most cases, the introduced resistance in these cultivars was broken within several years due to the development (via mutations) and natural selection of resistance-breaking viral strains. Among the numerous unidentified, empirically selected resistance genes, a few are distinguished as relatively stable single dominant genes: the TMV-resistant Tm family (Tm-2 and Tm-2a) introduced into tomato from *Lycopersicon peruvianum*,[277] the L(1-4) gene line of pepper, also protecting against TMV,[278] the Nb and Nx genes in potato, and the Rx gene of potato and other *Solanum* spp.,[279] protecting against PVX (Rx may provide protection against other, unrelated viruses as well[280]). Having been extensively investigated as early as the 1930s, the TMV-resistance-conferring N gene of *Nicotiana* species is the most studied of these genes. The N gene has been transferred from *N. glutinosa* to several tobacco cultivars.[281] It is an extremely stable gene and, until recently, none of the many natural and artificial TMV mutants could overcome N-gene-mediated resistance. However, a resistance-breaking TMV isolate has been reported, and the N-mediated resistance mapped to the gene of the 126-kDa protein, a member of the replicase-associated genes of TMV[282] (this may explain its stability, since most replicase-mutated viruses would probably be lethal). A similar gene of *N. sylvestris*, the N' gene, conferred resistance to only some TMV strains. It was inferred that the N' gene had evolved from the N gene,[248] but this is not very likely, since the N' resistance-breaking virus locus was mapped to the gene of the TMV coat protein.[59,283]

Many of these genes exert their protective influence by promoting viral localization, which in most cases is associated with tissue necrotization. The Rx gene is an example of a possible non-necrotizing, replication-blocking gene which may generally protect plants from viral infection.[280] These and similar genes are potential candidates for inducing resistance in transgenic plants devoid of the resistance gene. However, since until recently (see below) no sequence information and no protein product were known for any of them, their physical nature and mode of action remained an enigma.

The development of gene-tagging technology by restricted fragment length polymorphism (RFLP), random amplified polymorphic DNA (RAPD), transposons, and similar methods[284-286] established molecular anchors within or near the gene, from which methods such as chromosome walking would enable the identification and characterization of the gene itself. To date, this has been successful for only one of the above-mentioned genes, the N gene.[287] Indeed, introduction of the N gene into a sensitive tobacco cultivar resulted in a resistant transgenic plant.

Other stable resistance genes may be exploited once they are isolated and characterized. The Tm-2a gene in tomato is a good example, and chromosome walking from RFLP and RAPD markers is already under way. Another example is the Rx gene of potato. Additional information on that gene and its mode of action may lead to the deliberate manipulation of this gene, or Rx-derived metabolic pathways, for the construction of virus-resistant plants.

Expression of a whole set of genes is stimulated in plants following pathogen infections and other stresses, such as osmotic or heat shocks (for review see Carr and Klessig[288]). The products of these genes have been termed "pathogenesis-related" (PR) proteins. PR proteins are associated with general defense mechanisms: lignin deposition, appearance of phytoalexins, necrotization, etc. Several of them have been found to be hydrolytic enzymes such as chitinases and glucanases, and two were recognized by human interferon antibodies.[260] The expression of many of these genes is stimulated in tobacco by TMV infection, particularly in N-gene-carrying cultivars. It was hoped that some of these genes could be used as elicitors for virus-resistant transgenic plants — but they failed to engender resistance on the transgenes.[289,290] The N gene may be viewed as a master regulatory gene (as can be inferred from its sequence homology to other stimulating genes[287]). It is very possible that TMV infection stimulates expression of the N gene, which in turn may stimulate the expression of a whole array of genes. Thus a number of pathways may be induced by TMV infection, not all of which necessarily play any role in virus resistance — but some may. As the N gene has been isolated and characterized, it is now possible to carry out direct studies of the mechanisms of N-gene-induced activities and, with a better understanding, come back to PR-derived resistance.

Recently a model has been put forward suggesting that the low steady-state level of defense-mechanism proteins in normal, noninfected plants is due to ubiquitin-mediated protein degradation rather than low expression. Transgenic plants expressing a ubiquitin mutant that interferes with ubiquitin-dependent protein degradation exhibited resistance to pathogen infection, including some resistance to TMV. These transgenes were also phenotypically similar to resistant plants in their development of necrotic lesions.[291] While it is still premature to draw any conclusion about the precise mechanism of this type of resistance, a new type of protection derived from a host gene has been demonstrated.

It should be noted that increasing the copy number of an existing gene may bring about silencing of the endogenous as well as introduced genes, as discussed above for co-suppression. Thus, every cloning of a host gene should be considered with caution. The presence of a plant-virus inhibitor in extracts of *Phytolacca* spp. (pokeweed) was documented as early as 1914 (cited in Bawden[292]). The *Phytolacca* antiviral protein (PAP) was found to consist of three forms, active against a range of viruses,[293] including HIV.[294] PAP was found to be a ribosomal binding protein, deglycosylating a specific base on eukaryotic 28S rRNA, thus preventing binding of elongation factor 2 and arresting protein synthesis. It is therefore an inherent phytotoxin. PAP is localized in the cell wall,[295] from where it cannot exert its toxicity. However, it is easily extractable and is believed to enter the cytoplasm whenever the cell wall is damaged. PAP has recently been cloned into plants.[296] Transgenic plants expressing over 10 ng PAP per milligram protein were stunted, mottled, or even sterile, due to its phytotoxic effect. Plants expressing less than 5 ng PAP per milligram appeared normal. These plants (tobacco, *N. benthamiana*, and potato) were protected from PVX, CMV, and PVY (as well as aphid-inoculated PVY). This approach offers the possibility of developing broad-spectrum resistance by cloning a single gene. In his review, Wilson[6] cites some unpublished data in which antisense RNA of suicide genes (pap, ricin, bacterial enterotoxin) were cloned into plants with a minus-strand viral promoter of a subgenomic viral RNA at their 3' end. Infection with the cognate virus should promote expression of the toxin and selectively kill only the infected cells. The potential dangers of such an approach are also mentioned: the possibility that sense toxin mRNA would be transcribed from a plant promoter, resulting in nonselective suicide of the whole plant.

2.10 Is There a Risk in Releasing Virus-Resistant Transgenic Plants to the Environment?

De Zoeten[297] was the first to voice a word of caution. He suggested that new virus "varieties" could potentially arise from releasing artificially protected transgenic plants into the environment and exposing them to inoculation with viruses. He pointed out two possible major routes of generating potentially hazardous new viral genotypes and phenotypes: transencapsidation and RNA recombination.

Heterologous encapsidation occurs in nature.[298,299] The effect of coat protein on insect transmissibility has been recorded[298] and studied in more detail in the case of potyviruses.[300,301] A small modification in the DAG motif of a potyvirus coat protein changes the virus from aphid transmissible to nontransmissible,[300] or vice versa.[301] As mentioned above, coat-protein-mediated protection usually covers a very narrow range (i.e., it is only effective against the virus of origin of the transgene), and infection of the transgenic plant by a not-so-distant cognate virus is quite possible. This virus may be transencapsidated by the endogenous coat protein and its insect transmissibility characteristic thus altered. This has, in fact, been experimentally demonstrated.[148,302]

Heteroencapsidation has also been demonstrated between the unrelated (but insect transmissible) viruses CMV and AlMV.[303] It is difficult to assess whether transencapsidation poses a real environmental hazard. Conclusions can be drawn only through experience. As cited above, greenhouse and field experiments are quite widespread these days, and undesirable traits have not yet been reported. However, true evaluation is time dependent, and as only a short time has elapsed since the first field tests, the question remains open. A priori, however, transencapsidation does not seem to pose a real lasting hazard, since the new phenotypes it creates last a single generation, with further generations reverting to the original viral phenotype.

Viral RNA recombination is quite common and has been implicated in RNA virus evolution.[304,305] RNA recombination has also been reported for plant viruses.[306,313] Recently, a recombination event between a viral RNA and transcripts of a transgenic plant has also been reported.[314] RNA recombination results in new genotypes which may remain stable and be passed on to generations down the line. However, the only true risk assessment is the test of time, and so far no deleterious effects have been reported from the field. Until time proves differently, theoretical evaluations may provide a clue. Falk and Bruening[315] analyzed the situation and found little foreseeable risk to agriculture from RNA recombination between trangenes and viruses. They argue that RNA recombination is an infrequent event, occurring far less often than normal mutation rates, that the recombinant would only survive under a considerable selective pressure and then only in the unlikely event that it is more viable than the competing natural viruses. Natural mixed infections readily occur in nature and provide better opportunities for RNA recombination than transgene-virus recombination. They also point out that conventional breeding poses similar risks, as in many cases new virulent viral strains emerge as a result of the selective pressure brought in by the introduced resistance gene. Yet plant breeding is not being abandoned because its economic benefit to agriculture outweighs the losses incurred by sticking with natural, sensitive plant varieties.

REFERENCES

1. Beachy, R. N., Loesch-Fries, S., and Tumer, E., Coat protein-mediated resistance against virus infection, *Annu. Rev. Phytopathol.*, 28, 451, 1990.
2. Nejidat, A., Clarck, W. G., and Beachy, R. N., Engineered resistance against plant virus diseases, *Physiol. Plant.*, 80, 662, 1990.
3. Beachy, R. N., Coat protein mediated resistance in transgenic plants, in *Viral Genes and Plant Pathogenesis*, Pirone, T. P. and Shaw, G. C., Eds., Springer-Verlag, New York, 1990, 13.
4. Clarck, W. G., Register, J., III, and Beachy, R. N., Engineering virus resistance in transgenic plants, in *Plant Biology*, Vol. 11, Alan R. Liss, New York, 1990, 273.
5. Nelson, R. S., Powel, P. A., and Beachy, R. N., Coat protein mediated protection against virus infection, in *Genetic Engineering in Crop Plants*, Lycett, G. W. and Greirson, D., Eds., Butterworths, Boston, 1990, 13.
6. Wilson, T. M. A., Strategies to protect crop plants against viruses: pathogen-derived resistance blossoms, *Proc. Natl. Acad. Sci. U.S.A.*, 90, 3134, 1993.
7. Sturtevant, A. P. and Beachy, R. N., Virus resistance in transgenic plants: coat protein mediated resistance, in *Transgenic Plants Fundamentals and Applications*, Marcel Dekker, New York, 1993, 93.
8. Fitchen, J. H. and Beachy, R. N., Genetically engineered protection against viruses in transgenic plants, *Annu. Rev. Microbiol.*, 47, 739, 1993.
9. Beachy, R. N., Virus resistance through expression of coat protein genes, in *Biotechnology in Plant Disease Control*, Chet, I., Ed., Wiley-Liss, New York, 1993, 89.
10. Baulcombe, D., Novel strategies for engineering resistance in plants, *Curr. Opin. Biotechnol.*, 5, 117, 1994.
11. Powell-Abel, P., Nelson, R. S., De, B., Hoffmann, N., Rogers, S. G., Fraley, R. T., and Beachy, R. N., Delay of disease development in transgenic plants that express the tobacco mosaic virus coat protein gene, *Science*, 232, 738, 1986.
12. Nelson, R. S.,, Powell-Abel, P., and Beachy, R. N., Lesions and virus accumulation in inoculated transgenic tobacco plants expressing the coat protein gene of tobacco mosaic virus, *Virology*, 158, 126, 1987.
13. Nelson, R. S., McCormick, S. M., Delannay, X., Dube', P., Layton, J., Anderson, E. J., Kaniewska, M., Proksch, R. K., Horsch, R., Rogers, S. G., Fraley, R. T., and Beachy, R. N., Virus tolerance, plant growth, and field performance of transgenic tomato plants expressing coat protein from tobacco mosaic virus, *Biotechnology*, 6, 403, 1988.
14. Nejidat, A. and Beachy, R. N., Transgenic tobacco plant expressing a coat protein gene of tobacco mosaic virus are resistant to some other tobamoviruses, *Mol. Plant Microbe Interact.*, 3, 247, 1990.
15. Loesch-Fries, L. S., Merlo, D., Zinnen, T., Burhop, L., Hill, K., Jarvis, N., Nelson, S., and Halk, E., Expression of alfalfa mosaic virus RNA 4 in transgenic plant confers virus resistance, *EMBO J.*, 6, 1845, 1987.
16. Kallerhoff, J., Perez, P., Bouzoubaa, S., Ben Tahar, S., and Perret, J., Beet necrotic yellow vein virus coat protein-mediated protection in sugarbeet (*Beta vulgaris* L.) protoplasts, *Plant Cell Rep.*, 9, 224, 1990.

17. Quemada, H. D., Gonsalves, D., and Slighton, J. L., Expression of coat protein from cucumber mosaic virus strain C in tobacco: protection against infection by CMV strains transmitted mechanically or by aphids, *Phytopathology*, 81, 794, 1991.
18. Namba, S., Ling , K., Gonsalves, C., Gonsalves, D., and Slighton, J. L., Expression of a gene encoding the coat protein of cucumber mosaic virus (CMV) strain WL appears to provide protection to tobacco plants against infection by several different CMV strains, *Gene*, 107, 181, 1991.
19. Okuno, T., Nakayama, M., Yoshida, S., Furusawa, I., and Komiya, T., Comparative susceptibility of transgenic tobacco plants and protoplasts expressing the coat protein gene of cucumber mosaic virus to infection with virions and RNA, *Phytopathology*, 83, 542, 1993.
20. Hemenway, C., Fang, R. X., Kaniewski, W. K., Chua, N.-H., and Tumer, N. E., Analysis of the mechanism of protection in transgenic plants expressing the potato virus X coat protein or its antisense RNA, *EMBO J.*, 7, 1273, 1988.
21. Hoekema, A., Huisman, M. J., Molendijk, L., Van Den Elzen, P. J. M., and Cornelissen, B. J. C., The genetic engineering of two commercial potato cultivars for resistance to potato virus X, *Biotechnology*, 7, 273, 1989.
22. Lawson, C., Kaniewski, W., Haley, L., Rozman, R., Newell, C., Sanders, P., and Tumer, N. E., Engineering resistance to mixed virus infection in a commercial potato cultivar: resistance to potato virus X and potato virus Y in transgenic Russet Burbank, *Biotechnology*, 8, 127, 1990.
23. Ploeg, A. T., Mathis, A., Bol, J. F., Brown, D. J. F., and Robinson, D. J., Susceptibility of transgenic tobacco plants expressing tobacco rattle virus coat protein to nematode-transmitted and mechanically inoculated tobacco rattle virus, *J. Gen. Virol.*, 74, 2709, 1993.
24. Goldbach, R. W. and De Haan, P., Prospects of engineered forms of resistance against tomato spotted wilt virus, *Semin. Virol.*, 4, 381, 1993.
25. Brault, V., Candresse, T., Legall, O., Delbos, R. P., Lanneau, M., and Dunez, J., Genetically engineered resistance against grapevine chrome mosaic nepovirus, *Plant Mol. Biol.*, 21, 89, 1993.
26. Walter, B., Control of grapevine fanleaf nepovirus: progress in detection, cross protection and coat protein mediated resistance, in Proceedings of the Israeli-French Binational Symposium of Plant Virology, Paris, 1993, abstract.
27. Stark, D. M. and Beachy, R. N., Protection against potyvirus infection in transgenic plants: evidence for broad spectrum resistance, *Biotechnology*, 7, 1257, 1989.
28. Ling, K., Namba, S., Gonsalves, C., Slighton, J. L., and Gonsalves, D., Protection against detrimental effects of potyvirus infection in transgenic tobacco plants expressing the papaya ringspot virus coat protein gene, *Biotechnology*, 9, 752, 1991.
29. Namba, S., Ling, K., Gonsalves, C., Slighton, J. L., and Gonsalves, D., Protection of transgenic plants expressing the coat protein gene of watermelon mosaic virus II or zucchini yellow mosaic virus against 6 potyviruses, *Phytopathology*, 82, 940, 1992.
30. Regner, F., da Camara Machado, A., da Camara Machado, M. L., Steinkelner, H., Mattanovich, D., Hanzer, V., Weiss, H., and Katinger, H., Coat protein mediated resistance to plum pox virus in *Nicotiana clevelandii* and *benthamiana*, *Plant Cell Rep.*, 11, 30, 1992.
31. Ravelonandro, M., Monsion, M., Delbos, R., and Dunez, J., Variable resistance to plum pox virus and potato virus Y infection in *Nicotiana* plants expressing plum pox virus coat protein, *Plant Sci.*, 91, 157, 1993.
32. Kunik, T., Salomon, R., Zamir, D., Navot, N., Zeidan, M., Michelson, I., Gafni, Y., and Czosnek, H., Transgenic tomato plants expressing the tomato yellow leaf curl virus capsid protein are resistant to virus, *Biotechnology*, 12, 500, 1994.
33. Kaniewski, W., Lawson, C., Sammons, B., Haley, L., Hart, J., Delannay, X., and Tumer, N. E., Field resistance of transgenic Russel Burbank potato to effects of infection by potato virus X and potato virus Y, *Biotechnology*, 8, 750, 1990.

34. Kaniewski, W. K. and Thomas, P. E., Field testing of virus resistant transgenic plants, *Semin. Virol.*, 4, 389, 1993.

35. Sanders, P. R., Sammons, B., Kaniewski, W. K., Haley, L., Layton, J., LaValle, B. J., Delannay, X., and Tumer, N. E., Field resistance of transgenic tomatoes expressing the tobacco mosaic virus or tomato mosaic virus coat protein genes, *Phytopathology*, 82, 683, 1992.

36. Gonsalves, D., Chee, P., Providenti, R., Seem, R., and Slighton, J. L., Comparison of coat protein mediated and genetically-derived resistance in cucumbers to infection by cucumber mosaic virus under field conditions with natural challenge inoculations by vectors, *Biotechnology*, 10, 1562, 1993.

37. Jongedijk, E., Huisman, M. J., and Cornelissen, B. J. C., Agronomic performance and field resistance of genetically modified virus-resistant potato plants, *Semin. Virol.* 4, 407, 1993.

38. Yie, Y., Zhao, F., Zhao, S. Z., Liu, Y. Z., and Tien, P., High resistance to cucumber mosaic virus conferred by satellite RNA and coat protein in transgenic commercial tobacco cultivar g-140, *Mol. Plant Microbe Interact.*, 5, 460, 1992.

39. Van der Vlugt, R. A. A., and Goldbach, B. W., Tobacco plants transformed with the potato virus YN coat protein gene are protected against different PVY isolates and against aphid mediated infection, *Transgenic Res.*, 2, 109, 1993.

40. Lindbo, J. A. and Dougherty, W. G., Pathogen-derived resistance to a potyvirus: immune and resistant phenotypes in transgenic tobacco expressing altered forms of a potyvirus coat protein nucleotide sequence, *Mol. Plant Microbe Interact.*, 5, 144, 1992.

41. Register, J. C., III and Beachy, R. N., Resistance to TMV in transgenic plants results from interference with an early event in infection, *Virology*, 166, 524, 1988.

42. Wilson, T. M. A., Cotranslational disassembly increases the effeciency of expression of TMV-RNA in wheat germ cell-free extracts, *Virology*, 138, 353, 1984.

43. Shaw, J. G., Plaskitt, K. A., and Wilson, T. M. A., Evidence that tobacco mosaic virus particles disassemble cotranslationally *in vivo*, *Virology*, 148, 326, 1986.

44. Osbourn, J. K., Watts, J. W., Beachy, R. N., and Wilson, T. M. A., Evidence that nucleocapsid disassembly and a later step in virus replication are inhibited in transgenic tobacco protoplasts expressing TMV coat protein, *Virology*, 172, 370, 1989.

45. Osbourn, J. K., Plaskitt, K. A., Watts, J. W., and Wilson, T. M. A., Tobacco mosaic virus coat protein and reporter gene transcript containing the TMV origin-of-assembly sequence do not interact in double-transgenic tobacco plants: implication for coat protein-mediated protection, *Mol. Plant Microbe Interact.*, 2, 340, 1989.

46. Reimann-Philip, U. and Beachy, R. N., The mechanism(s) of coat protein-mediated resistance against tobacco mosaic virus, *Semin. Virol.*, 4, 349, 1993.

47. Hemenway, C., Haley, L., Kaniewski, W. K., Lawson, E. C., O'Connell, K. M., Sanders, P. R., Thomas, P. E., and Tumer, N. E., Genetically engineered resistance: transgenic plants, in *Plant Viruses*, Vol. 2, Pathology, Mandahar, C. L., Ed., CRC Press, Boca Raton, FL, 1990, 347.

48. MacKenzie, D. J., Tremaine, J. H., and McPherson, J., Genetically engineered resistance to potato virus S in potato cultivar Russet Burbank, *Mol. Plant Microbe Interact.*, 4, 95, 1991.

49. Bertioli, D. J., Cooper, J. I., Edwards, M. L., and Hawes, W. S., Arabis mosaic nepovirus coat protein in transgenic tobacco lessens disease severity and virus replication, *Ann. Appl. Biol.*, 120, 47, 1992.

50. Tumer, N. E., O'Connell, K. M., Nelson, R. S., Sanders, P. R., Beachy, R. N., Fraley, R. T., and Shah, D. M., Expression of alfalfa mosaic virus coat protein gene confers cross protection in transgenic tobacco and tomato plants, *EMBO J.*, 6, 1181, 1987.

51. Ward, R., Strand, M., and Valentine, R. C., Translational repression of f2 protein synthesis, *Biochem. Biophys. Res. Commun.*, 30, 310, 1968.

52. Galinski, M. S., Paramyxoviridae: transcription and replication, *Adv. Virus Res.*, 39, 129, 1991.
53. Gielen, J. J. L., de Haan, P., Kool, A. J., Peters, D., van Grinsven, M. Q. J. M., and Goldbach, R. W., Engineered resistance to tomato spotted wilt virus, a negative-stranded RNA virus, *Biotechnology*, 9, 1363, 1991.
54. Bol, J. F., Van Vloten-Doting, L., and Jaspars, E. M. J. A., Functional equivalence of top component RNA and coat protein in the initiation of infection by alfalfa mosaic virus, *Virology*, 46, 73, 1971.
55. Tumer, N. E., Kaniewski, W., Haley, L., Gehrke, L., Lodge, J. K., and Sanders, P. R., The second amino acid of alfalfa mosaic virus coat protein is critical for coat protein mediated protection, *Proc. Natl. Acad. Sci. U.S.A.*, 88, 2331, 1991.
56. Powell, P. A., Sanders, P. R., Tumer, N., Fraley, R. T., and Beachy, R. N., Protection against tobacco mosaic virus infection in transgenic plants requires accumulation of coat protein rather than coat protein RNA sequences, *Virology*, 175, 124, 1990.
57. Van Dun, C. M. P., Bol, J. F., and van Vloten-Doting, L., Expression of alfalfa mosaic virus and tobacco rattle virus coat protein genes in plants, *Virology*, 159, 299, 1987.
58. Mackenzie, D. J. and Ellis, P. J., Resistance to tomato spotted wilt virus infection in transgenic tobacco expressing the viral nucleocapsid gene, *Mol. Plant Microbe Interact.*, 5, 34, 1992.
59. van Dun, C. M. P., Overduin, B., van Vloten-Doting, L., and Bol, J. F., Transgenic tobacco expressing tobacco streak virus or mutated alfalfa mosaic virus coat protein does not cross-protect against alfalfa mosaic virus infection, *Virology*, 164, 383, 1988.
60. Dinant, S., Blaise, F., Kusiak, C., Astiermanifacier, S., and Albouy, J., Heterologous resistance to potato virus Y in transgenic tobacco plants expressing the coat protein gene of lettuce mosaic potyvirus, *Phytopathology*, 83, 818, 1993.
61. Anderson, E. J., Stark, D. M., Nelson, R. S., Powell, P. A., Tumer, N. E., and Beachy, R. N., Transgenic plants that express the coat protein of tobacco mosaic virus or alfalfa mosaic virus interfere with disease development of non-related viruses, *Phytopathology*, 79, 1284, 1989.
62. Lindbo, J. A. and Dougherty, W. G., Untranslatable transcripts of the tobacco etch virus coat protein gene sequence can interfere with tobacco etch virus replication in transgenic plants and protoplasts, *Virology*, 189, 725, 1992.
63. Lindbo, J. A. and Dougherty, W. G., Pathogen-derived resistance to a potyvirus — immune and resistant phenotypes in transgenic tobacco expressing altered forms of a potyvirus coat protein nucleotide sequence, *Mol. Plant Microbe Interact.*, 5, 144, 1992.
64. Van der Vlugt, R. A. A., Ruiter, R. K., and Goldbach, R. W., Evidence for sense RNA-mediated protection to PVYN in tobacco plants transformed with the viral coat protein cistron, *Plant Mol. Biol.*, 20, 631, 1992.
65. Farinelli, L. and Malone, P., Coat protein gene-mediated resistance to potato virus Y in tobacco — examination of the resistance mechanisms — is the transgenic coat protein required for protection?, *Mol. Plant Microbe Interact.*, 6, 284, 1993.
66. Lindbo, J. A., Silva-Rosales, L., and Dougherty, W. G., Pathogen-derived resistance to potyviruses: working, but why?, *Semin. Virol.* 4, 369, 1993.
67. Lindbo, J. A., Silva-Rosales, L., Proebsting, W. M., and Dougherty, W. G., Induction of highly specific antiviral state in transgenic plants: implication for regulation of gene expression and virus resistance, *Plant Cell*, 5, 1749, 1993.
68. De Haan, P., Gielen, J. J. L., Prins, M., van Grinsven, M. Q. J. M., and Goldbach, R. W., Characterization of RNA-mediated resistance to tomato spotted wilt virus in transgenic tobacco plants, *Biotechnology*, 10, 1133, 1992.
69. Goldbach, R. W. and De Haan, P., Prospects of engineered forms of resistance against tomato spotted wilt virus, *Semin. Virol.*, 4, 381, 1993.

70. Pang, S. Z., Slightom, J. L., and Gonsalves, D., Different mechanisms protect transgenic tobacco against tomato spotted wilt and impatiens necrotic spot topoviruses, *Biotechnology*, 11, 819, 1993.

71. Gelvin, S. B., Karcher, S. J., and DiRita, V. J., Methylation of the T-DNA in *Agrobacterium tumefaciens* and in several crown gall tumors, *Nucleic Acids Res.*, 11, 159, 1983.

72. Meyer, P., Heidmann, I., Forkmann, G., and Saedler, H., A new petunia flower colour generated by transformation of a mutant with a maize gene, *Nature*, 330, 677, 1987.

73. Linn, F., Haidmann, I., Saedler, H., and Meyer, P., Epigenetic changes in the expression of the maize AI gene in *Petunia hybrida*: role of numbers of integrated gene copies and state of methylation, *Mol. Gen. Genet.*, 222, 329, 1990.

74. Van der Krol, A. R., Mur, L. A., Beld, M., Mol, J. N. M., and Stuitje, A. R., Flavonoid genes in petunia: addition of a limited number of gene copies may lead to a suppression of gene expression, *Plant Cell*, 2, 291, 1990.

75. Fujiwara, T., Lessard, P. A., and Beachy, R. N., Inactivation of the nopaline synthase gene by double transformation: reactivation by segregation of the induced DNA, *Plant Cell Rep.*, 12, 133, 1993.

76. Kooter, J. M. and Mol, J. N. M., Trans inactivation of gene expression in plants. *Curr. Opin. Biotechnol.*, 4, 166, 1993.

77. Finnegan, J. and McElroy, D., Transgenic inactivation: plants fight back!, *Biotechnology*, 12, 883, 1994.

78. Grierson, D., Fray, R. G., Hamilton, A. J., Smith, C. J. S., and Watson, C. F., Does co-suppression of sense genes in transgenic plants involve antisense RNA?, *Trends Biotechnol.*, 4, 122, 1991.

79. Nelson, R. S., Roth, D. A., and Johnson, J. D., Tobacco mosaic virus infection of transgenic *Nicotiana tabacum* plants is inhibited by antisense constructs directed at the 5′ region of viral RNA, *Gene*, 127, 227, 1993.

80. Dougherty, W. G., Lindbo, J. A., Smith, H. A., Parks, T. D., Swaney, S., and Proebsting, W. M., RNA-mediated virus resistance in transgenic plants: exploitation of a cellular pathway possibly involved in RNA degradation, *Mol. Plant Microbe Interact.*, 7, 544, 1994.

81. Knorr, D. A. and Dawson, W. O., A point mutation in the tobacco mosaic virus capsid protein gene induces hypersensitivity in *Nicotiana sylvestris*, *Proc. Natl. Acad. Sci. U.S.A.*, 85, 170, 1988.

82. Padgett, H. S. and Beachy, R. N., Analysis of a tobacco mosaic virus strain capable of overcoming N gene mediated resistance, *Plant Cell*, 5, 577, 1993.

83. Wisnieski, L. A., Powell, P. A., Nelson, R. S., and Beachy, R. N., Local and systemic movement of tobacco mosaic virus (TMV) in tobacco plants that express the TMV coat protein gene, *Plant Cell*, 2, 559, 1990.

84. Kohm, B. A., Goulden, M. G., Gilbert, J. E., Kavanagh, T. A., and Baulcombe, D. C., A potato virus X resistance gene mediates an induced, nonspecific resistance in protoplasts, *Plant Cell*, 5, 913, 1993.

85. Nakajima, M., Hayakawa, T., Nakamura, I., and Suzuki, M., Protection against cucumber mosaic virus (CMV) strains-o and strains-y and chrysanthemum mild mottle virus in transgenic tobacco plants expressing CMV-o coat protein, *J. Gen. Virol.*, 74, 319, 1993.

86. Deom, C. M., Oliver, M. J., and Beachy, R. N., The 30-kilodalton gene product of tobacco mosaic virus potentiates virus movement, *Science*, 237, 384, 1987.

87. Meshi, T., Watanaba, Y., Saito, T., Sugimoto, A., Maeda, T., and Okada, Y., Function of the 30kD protein of tobacco mosaic virus: involvement in cell-to-cell movement and dispensibility for replication, *EMBO J.*, 6, 2557, 1987.

88. Wellink, J. and van Kammen, A., Cell-to-cell transport of cowpea mosaic virus requires both the 58K/48K proteins and the capsid proteins, *J. Gen. Virol.*, 70, 2279, 1989.

89. De Jong, W. and Ahlquist, P., A hybrid plant RNA virus made by transferring the noncapsid movement protein from a rod-shaped to icosahedral virus is competent for systemic infection, *Proc. Natl. Acad. Sci. U.S.A.*, 89, 6808, 1992.
90. Osman, T. A. M., Hayes, R. J., and Buck, K. W., Cooperative binding of the red clover necrotic mosaic virus movement protein to single-stranded nucleic acids, *J. Gen. Virol.*, 73, 223, 1992.
91. Shoumaker, F., Erny, C., Berna, A., Godefroy-Colburn, T., and Stussi-Garaud, C., Nucleic acid binding properties of alfalfa mosaic virus movement protein produced in yeast, *Virology*, 188, 896, 1992.
92. Ziegler-Graff, V., Guilford, P. J., and Baulcombe, D. C., Tobacco rattle virus RNA-1 29K gene product potentiates viral movement and also affects symptoms induction in tobacco, *Virology*, 182, 145, 1991.
93. Citovsky, V., Knorr, D., and Zambryski, P., Gene I, a potential movement locus of CaMV, encodes an RNA binding protein, *Proc. Natl. Acad. Sci. U.S.A.*, 88, 2476, 1991.
94. Atabekov, J. G. and Taliansky, M. E., Expression of plant virus-coded transport function by different viral genomes, *Adv. Virus Res.*, 38, 201, 1990.
95. Maule, A. J., Virus movement in infected plants, *Crit. Rev. Plant Sci.*, 9, 457, 1991.
96. Deom, C. M., Lapidot, M., and Beachy, R. N., Plant virus movement proteins, *Cell*, 69, 221, 1992.
97. Hull, R., The movement of viruses within plants, *Semin. Virol.*, 2, 89, 1992.
98. Lapidot, M. and Beachy, R. N., Transgenic plants: a tool to study intercellular transport, in *Transgenic Plants, Fundamentals and Applications*, Hiatt, A., Ed., Marcel Dekker, New York, 1993, 61.
99. Citovsky, V. and Zambryski, P., Transport of nucleic acid through membrane channels: snaking through small holes, *Annu. Rev. Microbiol.*, 47, 167, 1993.
100. Taliansky, M. E., Malishenko, S. I., Pshennikova, E. S., Kaplan, I. B., Ulanova, E. F., and Atabekov, J. G., Plant virus-specific transport function. I. Virus genetic control required for systemic spread, *Virology*, 122, 318, 1982.
101. Ohno, T., Takamatsu, N., Meshi, T., Okada, Y., Nishigichi, M., and Kiho, Y., Single amino acid substitution in 30K protein of TMV defective in virus transport function, *Virology*, 131, 255, 1983.
102. Zimmern, D. and Hunter, T., Point mutation in the 30kD open reading frame of TMV implicated in temperature sensitive assembly and local lesion spreading of mutant Ni2519, *EMBO J.*, 2, 1893, 1983.
103. Holt, C. A. and Beachy, R. N., *In-vivo* complementation of infectious transcripts from tobacco mosaic virus cDNAs in transgenic plants, *Virology*, 181, 109, 1991.
104. Tomenius, K., Claphan, D., and Meshi, T., Localization by immunogold cytochemistry of the virus-coded 30K protein in plasmodesmata of leaves infected with tobacco mosaic virus, *Virology*, 160, 363, 1987.
105. Wolf, S., Deom, C. M., Beachy, R. N., and Lucas, W. J., Movement protein of tobacco mosaic virus modifies plasmodesmata size exclusion limit, *Science*, 246, 377, 1989.
106. Deom, C. M., Schubert, K., Wolf, S., Holt, C. A., Lucas, W. J., and Beachy, R. N., Molecular characterization and biological function of the movement protein of tobacco mosaic virus in transgenic plants, *Proc. Natl. Acad. Sci. U.S.A.*, 87, 3284, 1990.
107. Atkins, D., Hull, R., Wells, B., Roberts, K., Moore, P., and Beachy, R. N., The tobacco mosaic virus 30K movement protein in transgenic tobacco plants is localized to plasmodesmata, *J. Gen. Virol.*, 72, 209, 1991.
108. Terry, B. R. and Robards, A. W., Hydrodynamic radius alone governs the mobility of molecules through plasmodesmata, *Planta*, 171, 145, 1987.
109. Citovsky, V., Knorr, D., Schuster, G., and Zambryski, P., The P30 movement protein of tobacco mosaic virus is a single strand nucleic acid binding protein, *Cell*, 60, 637, 1990.

110. Citovsky, V. and Zambryski, P., How do plant virus nucleic acids move through intercellular connections?, *BioEssays*, 13, 373, 1991.

111. Jackson, A. O., Petty, I. I. D., Jones, R. W., Edwards, M. C., and French, R., Analysis of barely stripe mosaic virus pathogenicity, *Semin. Virol.*, 2, 107, 1991.

112. Citovsky, V., Wong, M. L., Shaw, A. L., Prasad, B. V., and Zambryski, P., Visualization and characterization of tobacco mosaic virus movement protein binding to single-stranded nucleic acids, *Plant Cell*, 4, 397, 1992.

113. Malyshenko, S. I., Kondakova, O. A., Nazarova, J. V., Kaplan, I. B., Taliansky, M. E., and Atabekov, J. G., Reduction of tobacco mosaic virus accumulation in transgenic plants producing non-functional viral transport protein, *J. Gen. Virol.*, 74, 1149, 1993.

114. van Lent, J., Wellink, J., and Goldbach, R., Evidence for the involvement of the 58K and 48K proteins in the intercellular movement of cowpea mosaic virus, *J. Gen. Virol.*, 71, 219, 1990.

115. van Lent, J., Storms, M., van der Meer, F., Wellink, J., and Goldbach, R., Tubular structures involved in movement of cowpea mosaic virus are also formed in infected cowpea protoplasts, *J. Gen. Virol.*, 72, 2615, 1991.

116. Malyshenko, S. I., Kondakova, O. A., Taliansky, M. E., and Atabekov, J. G., Plant virus transport function: complementation by helper virus is non-specific, *J. Gen. Virol.*, 69, 407, 1988.

117. Taliansky, M. E., Malyshenko, S. I., Kaplan, I. B., Kondakova, O. A., and Atabekov, J. G., Production of tobacco mosaic virus (TMV) transport protein in transgenic plants is essential but insufficient for complementing foreign virus transport: a need for the full-length TMV genome or some other TMV-encoded product, *J. Gen. Virol.*, 73, 471, 1992.

118. Lapidot, M., Gafny, R., Ding, B., Wolf, S., Lucas, W. J., and Beachy, R. N., A dysfunctional movement protein of tobacco mosaic virus that partially modifies the plasmodesmata and limits virus spread in transgenic plants, *Plant J.*, 4, 959, 1993.

119. Young, N., Forney, J., and Zaitlin, M., Tobacco mosaic virus replicase and replicative structures, *J. Cell Sci.*, Suppl. 7, 177, 1987.

120. Lomonossoff, G. P., Virus resistance mediated by a nonstructural viral gene sequence, in *Transgenic Plants, Fundamentals and Applications*, Hiatt, A., Ed., Marcel Dekker, New York, 1993, 79.

121. Baulcombe, D. C., Replicase mediated resistance: a novel type of virus resistance in transgenic plants?, *Trends Microbiol.*, 2, 60, 1994.

122. Carr, J. P. and Zaitlin, M., Replicase-mediated resistance, *Semin. Virol.*, 4, 339, 1993.

123. Golemboski, D. B., Lomonossoff, G. P., and Zaitlin, M., Plants transformed with a tobacco mosaic virus nonstructural gene sequence are resistant to virus, *Proc. Natl. Acad. Sci. U.S.A.*, 87, 6311, 1990.

124. Donson, J., Kearney, C. M., Turpen, T. H., Khan, I. A., Kurath, G., Turpen, A. M., Jones, G. E., Dawson, W. O., and Lewandowski, D. J., Broad resistance to tobamoviruses is mediated by a modified tobacco mosaic virus replicase transgene, *Mol. Plant Microbe Interact.*, 6, 635, 1993.

125. Anderson, J. M., Palukaitis, P., and Zaitlin, M., A defective replicase gene induces resistance to cucumber mosaic virus in transgenic tobacco plants, *Proc. Natl. Acad. Sci. U.S.A.*, 89, 8759, 1992.

126. Hemenway, C. L. and Braun, C. J., Expression of amino terminal portion or full-length viral replicase genes in transgenic plants confers resistance to potato virus X infection, *Plant Cell*, 4, 735, 1992.

127. Longstaff, M., Brigneti, G., Boccard, F., Chapman, S., and Baulcombe, D. C., Extreme resistance to potato virus X infection in plants expressing a modified component of the putative viral replicase, *EMBO J.*, 12, 379, 1993.

128. Audy, P., Palukaitis, P., Slack, S. A., and Zaitlin, M., Replicase-mediated resistance to potato virus Y in transgenic plants, *Mol. Plant Microbe Interact.*, 7, 15, 1994.

129. MacFarlane, S. A. and Davis, J. W., Plants transformed with a region of the 201-kilodalton replicase gene from pea early browning virus RNA1 are resistant to virus infection, *Proc. Natl. Acad. Sci. U.S.A.*, 89, 5829, 1992.

130. Rubino, L., Lupo, R., and Russo, M., Resistance to cymbidium ringspot virus infection in transgenic plants expressing full-length viral replicase gene, *Mol. Plant Microbe Interact.*, 6, 729, 1993.

131. van Dun, C. M. P., van Vloten-Doting, L., and Bol, J. F., Expression of alfalfa mosaic virus cDNA 1 and 2 in transgenic tobacco plants, *Virology*, 163, 572, 1988.

132. Mori, M., Mise, K., Okuno, T., and Furusawa, I., Expression of brome mosaic virus-encoded replicase genes in transgenic tobacco plants, *J. Gen. Virol.*, 73, 169, 1992.

133. Sulzinski, M. A., Gabard, K. A., Palukaitis, P., and Zaitlin, M., Replication of tobacco mosaic virus. VIII. Characterization of a third subgenomic TMV-RNA, *Virology*, 145, 132, 1985.

134. Zelcer, A., Weaber, K. F., Balazes, E., and Zaitlin, M., The detection and characterization of viral-related double-stranded RNAs in tobacco mosaic virus-infected plants, *Virology*, 113, 417, 1981.

135. Carr, J. P., Marsh, L. E., Lomonossoff, G. P., Sekiya, M. E., and Zaitlin, M., Resistance to tobacco mosaic virus induced by the 54-kDa gene sequence requires expression of the 54-kDa protein, *Mol. Plant Microbe Interact.*, 5, 397, 1992.

136. Vardi, E., Sela, I., Edelbaum, O., Livneh, O., Kuznetsova, L., and Stram, Y., Plants transformed with a cistron of potato virus Y protease (NIa) are resistant to virus infection, *Proc. Natl. Acad. Sci. U.S.A.*, 90, 7513, 1993.

137. Maiti, J. B., Murphy, J. F., Shaw, J. G., and Hunt, A. G., Plants that express a potyvirus proteinase gene are resistant to virus infection, *Proc. Natl. Acad. Sci. U.S.A.*, 90, 6110, 1993.

138. Livneh, O., Edelbaum, O., Kuznetsova, L., Livne, B., Vardi, E., and Sela, I., Plants transformed with the first (nonstructural) three cistrons of potato virus Y are resistant to potato virus infection, *Transgenics*, 1994 (in press).

139. Allison, R., Johnston, R. E., and Dougherty, W. G., The nucleotide sequence of the coding region of tobacco etch virus genomic RNA: evidence for the synthesis of a single polyprotein, *Virology*, 154, 9, 1986.

140. Hellmann, G. M., Hiremath, S. T., Shaw, J. G., and Rhodas, R. E., Cistron mapping of tobacco vein mottling virus, *Virology*, 151, 159, 1986.

141. Robaglia, C., Durand-Tardif, M., Tronchey, M., Boudazin, G., Astier-Manifacier, S., and Casse-Delbart, F., Nucleotide sequence of potato virus Y (N strain) genomic RNA, *J. Gen. Virol.*, 70, 935, 1989.

142. Carrington, J. C., Cary, S. M., and Dougherty, W. G., Mutational analysis of tobacco etch virus polyprotein processing; cis and trans proteolytic activities of polyprotein containing the 49-kilodalton proteinase, *J. Virol.*, 62, 2313, 1988.

143. Hellmann, G. M., Shaw, J. C., and Rhodas, R. E., In vitro analysis of tobacco vein mottling virus NIa cistron: evidence for a virus-encoded protease, *Virology*, 163, 554, 1988.

144. Carrington, J. C., Cary, S. M., Park, T. D., and Dougherty, W. G., A second proteinase encoded by a plant potyvirus genome, *EMBO J.*, 8, 365, 1989.

145. Vechot, J., Koonin, E. V., and Carrington, J. C., The 35-kDa protein from the N-terminus of the potyviral polyprotein functions as a third virus-encoded proteinase, *Virology*, 185, 527, 1991.

146. Stram, Y., Chetsrony, A., Karchi, H., Karchi, M., Edelbaum, O., Vardi, E., Livneh, O., and Sela, I., Expression of the "helper component" protein of potato virus Y (PVY) in *E. coli*: possible involvement of a third protease, *Virus Genes*, 7, 151, 1993.

147. Vardi, E., Stram, Y., Livneh, O., Edelbaum, O., and Sela, I., Transgenic plants expressing various PVY genes: studies on the mechanism of protein processing and the induction of resistance by non-structural genes, in Proceedings of the Israeli-French Binational Symposium of Plant Virology, Paris, 1993, abstract.

148. Lecoq, H., Ravelolandro, M., Wipf-Scheibel, C., Monsion, M., Raccah, B., and Dunez, J., Aphid transmission of non-aphid-transmissible strain of zucchini yellow mosaic potyvirus from transgenic plants expressing the capsid protein of plum pox potyvirus, *Mol. Plant Microbe Interact.*, 6, 403, 1993.

149. Lergarve, T., Durand-Tardif, M., Casse-Delbart, F., and Robaglia, C., Search for new resistance genes derived from the potato virus Y genome, in Proceedings of the Israeli-French Binational Symposium of Plant Virology, Paris, 1993, abstract.

150. Robaglia, C., personal communication, 1994.

151. Murant, A. F. and Mayo, M. A., Satellite of plant viruses, *Annu. Rev. Plant Pathol.*, 20, 49, 1982.

152. Francki, R. I. B., Plant virus satellite, *Annu. Rev. Microbiol.*, 39, 151, 1986.

153. Yie, Y. and Tien, P., Plant virus satellite RNAs and their role in engineering resistance to virus diseases, *Semin. Virol.*, 4, 363, 1993.

154. Wu, G. S., Kaper, J. M., and Jaspars, E. M. J., Replication of cucumber mosaic virus satellite RNA in vitro by an RNA-dependent RNA polymerase from virus infected tobacco, *FEBS Lett.*, 292, 213, 1991.

155. Hayes, R. T., Tousch, D., Jacquemond, M., Pereira, V. C., Buck, K. W., and Tepfer, M., Complete replication of a satellite RNA in vitro by a purified RNA-dependent RNA polymerase, *J. Gen. Virol.*, 73, 1597, 1992.

156. Delmer, S. A., Rucker, D. G., Nooruddin, L., and De-Zotten, G. A., Replication of the satellite RNA of pea enation mosaic virus is controlled by RNA 2-encoded functions, *J. Gen. Virol.*, 75, 1399, 1994.

157. Liu, Y. Y. and Cooper, J. I., The multiplication in plants of arabis mosaic virus satellite RNA requires the encoded protein, *J. Gen. Virol.*, 74, 1471, 1993.

158. Hemmer, O., Oncino, C., and Fritsch, C., Efficient replication of the in vitro transcripts from cloned cDNA of tomato black ring virus satellite RNA requires the 48K satellite-encoded protein, *Virology*, 194, 800, 1993.

159. Liu, Y. Y., Cooper, J. I., Coates, D., and Bauer, G., Biologically active transcripts of a large satellite RNA from arabis mosaic nepovirus and the importance of the 5′ end sequences for its replication, *J. Gen. Virol.*, 72, 2867, 1991.

160. Hidaka, S. and Hanada, K., Structural features unique to a new 405-nucleotide satellite RNA of cucumber mosaic virus inducing tomato necrosis, *Virology*, 200, 806, 1994.

161. van Tol, H., Buzayan, J. M., and Bruening, G., Evidence for spontaneous circle formation in the replication of the satellite RNA of tobacco ringspot virus, *Virology*, 180, 23, 1991.

162. Hernandez, C., Daros, J. A., Elena, S. F., Moya, A., and Flores, R., The strands of both polarities of a small circular RNA from carnation self-cleave in vitro through alternative double- and single-hammerhead structures, *Nucleic Acids Res.*, 20, 6323, 1992.

163. Dzianott, A. M. and Bujarski, J. J., Derivation of an infectious viral RNA by autolytic cleavage of in vitro transcribed viral complementary DNA, *Proc. Natl. Acad. Sci. U.S.A.*, 86, 4823, 1989.

164. Kurath, G. and Palukaitis, P., Serial passage of infectious transcripts of a cucumber mosaic virus satellite RNA clone results in sequence heterogeneity, *Virology*, 176, 8, 1990.

165. Moriones, E., Fraile, A., and Garcia-Arenal, F., Host-associated selection of sequence variants from satellite RNA of cucumber mosaic virus, *Virology*, 184, 465, 1991.

166. Aranda, M. A., Fraile, A., and Garcia-Arenal, F., Genetic variability and evolution of the satellite RNA of cucumber mosaic virus during natural epidemic, *J. Virol.*, 67, 5896, 1993.

167. Collmer, C. W. and Howell, S. H., Role of satellite RNA in the expression of symptoms caused by plant viruses, *Annu. Rev. Phytopathol.*, 30, 419, 1992.

168. Kaper, J. M., Rapid synthesis of double-stranded cucumber mosaic virus-associated RNA 5: mechanism controlling viral pathogenesis?, *Biochem. Biophys. Res. Commun.*, 105, 1014, 1982.

169. Masuta, C., Suzuki, M., Matsuzaki, T., Honda, I., Kuwata, S., Takanami, Y., and Koiwai, A., Bright yellow chlorosis by cucumber mosaic virus Y satellite RNA is specifically induced without severe chloroplast damage, *Physiol. Mol. Plant Pathol.*, 42, 267, 1993.

170. Masuta, C. and Takanami, Y., Determination of sequence and structural requirements for pathogenecity of a cucumber mosaic virus satellite RNA (Y-satellite RNA), *Plant Cell*, 1, 1165, 1989.

171. Zhang, L., Kim, C. H., and Palukaitis, P., The chlorosis-induction domain of the satellite RNA of cucumber mosaic virus: identifying sequences that affect accumulation and the degree of chlorosis, *Mol. Plant Microbe Interact.*, 7, 208, 1994.

172. Kaper, J. M. and Waterworth, H. E., Cucumber mosaic virus associated RNA 5: causal agent for tomato necrosis, *Science*, 196, 429, 1977.

173. Kaper, J. M. and Tousignant, M. E., Viral satellite: parasitic nucleic acids capable of modulating disease expression, *Endeavour New Ser.*, 8, 194, 1984.

174. Devic, M., Jeagle, M., and Baulocomb, D., Symptom production on tobacco and tomato is determined by two distinct domains of the satellite RNA of cucumber mosaic virus (strain Y), *J. Gen. Virol.*, 70, 2765, 1989.

175. Sleat, D. E. and Palukaitis, P., Site-directed mutagenesis of a plant viral satellite RNA changes its type from ameliorative to necrogenic, *Proc. Natl. Acad. Sci. U.S.A.*, 87, 2946, 1990.

176. Kurath, G. and Palukaitis, P., Satellite RNA of cucumber mosaic virus: recombinants constructed in vitro reveal independent functional domains for chlorosis and necrosis in tomato, *Mol. Plant Microbe Interact.*, 2, 91, 1989.

177. Sleat, D. E., Zhang, L., and Palukaitis, P., Mapping determinants within cucumber mosaic virus and its satellite RNA for the induction of necrosis in tomato plants, *Mol. Plant Microbe Interact.*, 7, 189, 1994.

178. Sleat, D. E. and Palukaitis, P., Induction of tobacco chlorosis by certain cucumber mosaic virus satellite RNA is specific to subgroup II helper strains, *Virology*, 176, 292, 1990.

179. Collmer, C. W., Stenzler, L., Chen, X., Fay, N., Hacker, D., and Howell, S. H., Single amino acid change in the helicase domain of the putative RNA replicase of turnip crinkle virus alters symptom intensification by virulent satellites, *Proc. Natl. Acad. Sci. U.S.A.*, 89, 309, 1992.

180. Masuta, C., Suzuki, M., Kuwata, S., Takanami, Y., and Koiway, A., Yellow mosaic symptoms induced by satellite RNA of cucumber mosaic virus is regulated by a single incompletely dominant gene in wild *Nicotiana* species, *Phytopathology*, 83, 411, 1993.

181. Baulcomb, D., Saunder, G. R., Revan, M. W., Mayo, M. A., and Harrison, B. D., Expression of biologically active viral satellite RNA from the nuclear genome of transformed plants, *Nature*, 321, 446, 1986.

182. Gerlach, W. L., Llewellyn, D., and Haseloff, J., Construction of a plant disease resistance gene from the satellite RNA of tobacco ringspot virus, *Nature*, 328, 802, 1987.

183. Tousch, D., Jacquemond, M., and Tepfer, M., Replication of cucumber mosaic virus satellite RNA from negative-sense transcripts produced either in vitro or in transgenic plants, *J. Gen. Virol.*, 75, 1009, 1994.

184. Jacquemond, M., Amselem, J., and Tepfer, M., A gene coding for a monomeric form of cucumber mosaic virus satellite RNA confers tolerance to CMV, *Mol. Plant Microbe Interact.*, 1, 311, 1988.

185. Wu, S. X., Zhao, S. Z., Wang, G. J., and Tien, P., Transgenic tobacco plants with resistance to cucumber mosaic virus by expressing satellite cDNA, *Sci. Sin. B*, 9, 948, 1989.

186. Zhao, S. Z., Wang, X., Wang, G. J., and Tien, P., Transgenic tomato resistant to cucumber mosaic virus by expressing its monomer and dimer satellite cDNA, *Sci. Sin. B*, 7, 708, 1990.

187. Wu, G., Kang, L., and Tien, P., The effect of satellite RNA on cross-protection among cucumber mosaic virus strains, *Ann. Appl. Biol.*, 114, 489, 1989.

188. Tien, P. and Wu, G., Satellite RNA for the biocontrol of plant disease, *Adv. Virus Res.*, 39, 321, 1991.

189. Saito, Y., Komari, T., Masuta, C., Hayashi, Y., Kumashiro, T., and Takanami, Y., Cucumber mosaic virus-tolerant transgenic tomato plants expressing a satellite RNA, *Theor. Appl. Genet.*, 83, 679, 1992.

190. McGarvey, P. B., Montasser, M. S., and Kaper, J. M, Transgenic tomato plants expressing satellite RNA are tolerant to some strains of cucumber mosaic virus, *J. Am. Soc. Hort. Sci.*, 119, 642, 1994.

191. Touch, D., Jacquemond, M., and Tepfer, M., Transgenic tomato plants expressing a cucumber mosaic virus (CMV) satellite RNA gene: inoculation with CMV induces lethal necrosis, *C.R. Acad. Sci. Ser. III*, 311, 377, 1990.

192. McGarvey, P. B., Kaper, J. M., Avial-Rincon, M. J., Pena, L., and Diaz-Ruiz, J. R., Transformed tomato plants express a satellite RNA of cucumber mosaic virus and produce lethal necrosis upon infection with viral RNA, *Biochem. Biophys. Res. Commun.*, 170, 548, 1990.

193. Roux, L., Simon, A. E., and Holland, J. J., Effect of defective interfering viruses on virus replication and pathogenesis in vitro and in vivo, *Adv. Virus Res.*, 40, 181, 1991.

194. Kollar, A., Dalmay, T., and Burgyan, J., Defective interfering RNA-mediated resistance against cymbidium ringspot tombuvirus in transgenic plants, *Virology*, 193, 313, 1993.

195. Resende, R. de O., de Haan, P., van de Vossen, E., de Avila, A. C., Goldbach, R., and Peters, D., Defective interfering L RNA segments of tomato spotted wilt virus retain both viral genome termini and have extensive internal deletions, *J. Gen. Virol.*, 73, 2509, 1992.

196. Li, X. H., Heaton, L. A., Morris, T. J., and Simon, A. E., Turnip crinkle virus defective interfering RNAs intensify viral symptoms and are generated de novo, *Proc. Natl. Acad. Sci. U.S.A.*, 86, 9173, 1989.

197. Resende, R. de O., personal communication, 1995.

198. Pogue, G. P., Marsh, L. E., Connell, J. P., and Hall, T. C., Requirement for ICR-like sequences in the replication of brome mosaic virus genomic RNA, *Virology*, 188, 742, 1992.

199. Pogue, G. P. and Hall, T. C., The requirement for 5′ stem-loop structure in brome mosaic virus replication supports a new model for viral positive-strand RNA initiation, *J. Virol.*, 66, 674, 1992.

200. Raffo, A. J. and Dawson, W. O., Construction of tobacco mosaic virus subgenomic replicons that are replicated and spread systemically in tobacco plants, *Virology*, 184, 277, 1991.

201. Morch, M. D., Joshi, R. L., Denial, T. M., and Haenni, A. L., A new sense RNA approach to block viral RNA replication in vitro, *Nucleic Acids Res.*, 15, 4123, 1987.

202. Huntley, C. C. and Hall, T. C., Interference with brome mosaic virus replication by targetting the minus strand promoter, *J. Gen. Virol.*, 74, 2445, 1993.

203. He'le'ne, C. and Toulme, J. J., Specific regulation of gene expression by antisense, sense, and antigen nucleic acids, *Biochim. Biophys. Acta*, 1049, 99, 1990.

204. Sassone-Corsi, P., Wildeman, A., and Chambon, P., A trans-acting factor is responsible for the simian virus 40 enhancer activity, *Nature*, 313, 458, 1985.

205. Melton, D. A., Injected antisense RNA specifically block messenger RNA translation, *Proc. Natl. Acad. Sci. U.S.A.*, 82, 144, 1985.

206. Hirashima, A., Sauki, S., Mizuno, T., Houba-Herin, N., and Inouge, M., Artificial immune system against viral infection involving antisense RNA targeted to the 5′ terminal noncoding region of coliphage SP RNA, *J. Biochem.*, 106, 163, 1989.

207. Kasid, U., Pfeifer, A., Brennan, T., Beckett, M., Weichselbaum, R. R., Dritschilo, A., and Mark, G. E., Effect of antisense c-raf-1 on tumorgenicity and radiation sensitivity of a human squamous carcinoma, *Science*, 243, 1354, 1989.
208. Strickland, G., Huarte, J., Belin, D., Vassalli, A., Rickles, R. J., and Vassalli, J. D., Antisense RNA directed against the 3′ noncoding region prevents dormant mRNA activation in mouse oocytes, *Science*, 241, 680, 1988.
209. Rothstein, S. J., DiMaio, J., Strand, M., and Rice, D., Stable and heritable inhibition of the expression of nopaline synthase in tobacco expressing antisense RNA, *Proc. Natl. Acad. Sci. U.S.A.*, 84, 8439, 1987.
210. Krol, A. R., Lenting, P. E., Veenstra, J., van der Meer, I., Koes, R. E., Gerates, A. G. M., Mol, J. N. M., and Stuitje, A. R., An antisense chalcone synthase gene in transgenic plants inhibits flower pigmentation, *Nature*, 333, 866, 1988.
211. Rodermel, S. R., Abbott, M. S., and Bogorad, L., Nuclear-organelle interactions: nuclear antisense gene inhibits ribulose biphosphate carboxylase enzyme level in transformed tobacco plants, *Cell*, 55, 673, 1988.
212. Smith, C. J. S., Watson, C. F., Ray, J., Bird, C. R., Morris, P. C., Schuch, W., and Greirson, D., Antisense RNA inhibition of polygalacturonase gene expression in transgenic tomatoes, *Nature*, 334, 724, 1988.
213. Hoefgen, R., Axelsen, K. B., Kannangara, C. G., Schuettke, I., Pohlnez, H. D., Willmitzer, L., Grimm, B., and Von Westtstein, D., A visible marker for antisense mRNA expression in plants: inhibition of chlorophyll synthesis with a glutamate-1-semialdehyde aminotransferase antisense gene, *Proc. Natl. Acad. Sci. U.S.A.*, 91, 1726, 1994.
214. Palomares, R., Herrmann, R. G., and Oelmueller, R., Antisense RNA for components associated with the oxygen-evolving complex and the Rieske iron/sulfur protein of the tobacco thylakoid membrane suppresses accumulation of mRNA, but not of protein, *Planta*, 190, 305, 1993.
215. Temples, S. J., Knight, T. J., Unkefer, P. J., and Sengupta-Gopalan, C., Modulation of glutamine synthetase gene expression in tobacco by the introduction of an alfalfa glutamine synthetase gene in sense and antisense orientation: molecular and biochemical analysis, *Mol. Gen. Genet.*, 238, 315, 1993.
216. Oliver, M. J., Ferguson, D. L., Burke, J. J., and Velten, J., Inhibition of tobacco NADH-hydroxypyruvate reductase by expression of a heterologous antisense RNA derived from cucumber cDNA: implications for the mechanism of action of antisense RNAs, *Mol. Gen. Genet.*, 239, 425, 1993.
217. Cuozzo, M., O'Conell, K. M., Kaniewski, W., Fang, R., Chua, N. H., and Tumer, N. E., Viral protection in transgenic tobacco plants expressing the cucumber mosaic virus coat protein or its antisense RNA, *Biotechnology*, 6, 549, 1988.
218. Powell, P. A., Stark, D. M., Sanders, P. R., and Beachy, R. N., Protection against tobacco mosaic virus in transgenic plants that express tobacco mosaic virus antisense RNA, *Proc. Natl. Acad. Sci. U.S.A.*, 86, 6949, 1989.
219. Fang, G. and Grumet, R., Genetic engineering of potyvirus resistance using constructs derived from the zucchini yellow mosaic coat protein gene, *Mol. Plant Microbe Interact.*, 6, 358, 1993.
220. Kawchuk, L. M., Martin, R. R., and McPherson, J., Sense and antisense RNA-mediated resistance to potato leafroll virus in Russet Burbank potato plants, *Mol. Plant Microbe Interact.*, 4, 247, 1991.
221. Day, A. G., Bejarano, E. R., Buck, K. W., Burrell, M., and Lichtenstein, C. P., Expression of an antisense viral gene in transgenic tobacco confers resistance to the DNA virus tomato golden mosaic virus, *Proc. Natl. Acad. Sci. U.S.A.*, 88, 6721, 1991.
222. Bejarno, E. R. and Lichtenstein, C. P., Expression of TGMV antisense RNA in transgenic tobacco inhibits replication of BCTV but not ACMV geminiviruses, *Plant Mol. Biol.*, 24, 241, 1994.

223. Ceck, T. R. and Bass, B. L., Biological catalysis by RNA, *Annu. Rev. Biochem.*, 55, 599, 1986.

224. Symons, R. H., Self-cleavage of RNA in the replication of small pathogenes of plants and animals, *Trends Biochem. Sci.*, 14, 445, 1989.

225. Forster, A. C. and Symons, R. H., Self-cleavage of plus and minus RNAs of a virusoid and a structural model for the active sites, *Cell*, 49, 211, 1987.

226. Forster, A. C., Davis, C., Sheldon, C. C., Jeffries, A. C., and Symons, R. H., Self-cleaving viroid and newt RNAs may only be active as dimers, *Nature*, 334, 265, 1988.

227. Jeffries, A. C. and Symons, R .H., A catalytic 13-mer ribozyme, *Nucleic Acids Res.*, 17, 1371, 1989.

228. Haseloff, J. and Gerlach, W. L., A simple RNA enzyme with new and highly specific endoribonuclease activities, *Nature*, 334, 585, 1988.

229. Cotten, M., The in vivo application of ribozymes, *Trends Biotechnol.*, 8, 174, 1990.

230. Sarver, N., Cantin, E., Chang, P., Zaia, J., Ladine, P., Stephens, D., and Rossi, J. J., Ribozymes as potential anti-HIV therapeutic agents, *Science*, 247, 1222, 1990.

231. Park, S. G. and Hwang, Y. S., In vitro endonucleolytic cleavage of synthesized cucumber mosaic virus RNA by hammerhead ribozyme, *Agric. Chem. Biotechnol.*, 37, 56, 1994.

232. Edington, B. V., Dixon, R. A., and Nelson, R. S., Ribozymes: description and uses, in *Transgenic Plants, Fundamentals and Applications*, Hiatt, A., Ed., Marcel Dekker, New York, 1993, 301.

233. Mazzolini, L., Axelos, M., Lescure, N., and Yot, P., Assaying synthetic ribozymes in plants: high-level expression of a functional hammerhead structure fails to inhibit target gene activity in transiently transformed protoplasts, *Plant Mol. Biol.*, 20, 715, 1992.

234. Bennet, M. J. and Cullimore, J. V., Selective cleavage of closely-related mRNAs by synthetic ribozymes, *Nucleic Acids Res.*, 20, 831, 1992.

235. Lamb, J. W. and Hay, R. T., Ribozyme that cleave potato leafroll virus RNA within the coat protein and polymerase genes, *J. Gen. Virol.*, 71, 2257, 1990.

236. Plucktun, A., Antibody engineering, *Curr. Opin. Biotechnol.*, 2, 238, 1991.

237. Hiatt, A., Cafferkey, R., and Bowdish, K., Production of antibodies in transgenic plants, *Nature*, 342, 469, 1989.

238. Benvenuto, E., Ordas, R. J., Tavazza, R., Ancora, G., Biocca, S., Cattaneo, A., and Galeffi, P., "Phytoantibodies": a general vector for the expression of immunoglobulin domains in transgenic plants, *Plant Mol. Biol.*, 17, 865, 1991.

239. De Neve, M., De Loose, M., Jacobes, A., van Houdt, H. A., Kaluza, B., Weidle, U., van Montagu, M., and Depicker, A., Assembly of an antibody and its derived antibody fragment in *Nicotiana* and *Arabidopsis*, *Transgenic Res.*, 2, 227, 1993.

240. Owen, M., Gandecha, A., Cockburn, B., and Whitelam, G., Synthesis of a functional antiphytochrome single-chain Fv protein in transgenic tobacco, *Biotechnology*, 10, 790, 1992.

241. Tavladoraki, P., Benvenuto, E., Trinca, S., De Martinis, D., Cattaneo, A., and Galeffi, P., Transgenic plants expressing a functional single-chain Fv antibody are protected from virus attack, *Nature*, 366, 469, 1993.

242. Voss, A., Niersbach, M., Hain, R., Hirsch, H. J., Liao, Y., Kreuzaler, F., and Fischer, R., Reduced virus infectivity in *N. tabacum* secreting a TMV-specific full size antibody, *Plant Mol. Biol.* (in press).

243. Sela, I. and Applebaum, S. W., Occurrence of antiviral factor in virus-infected plants, *Virology*, 17, 543, 1962.

244. Sela, I., Harpaz, I., and Birk, Y., Seperation of a highly active antiviral factor from virus-infected plants, *Virology*, 22, 446, 1964.

245. Sela, I., Preparation and measurement of an antiviral protein found in tobacco cells after infection with tobacco mosaic virus, *Methods Enzymol.*, 119, 734, 1986.

246. Antignus, Y., Sela, I., and Harpaz, I., A phosphorous-containing fraction associated with antiviral activity in *Nicotiana* spp. carrying the gene for localization of infection, *Physiol. Plant Pathol.*, 6, 159, 1975.

247. Antignus, Y., Sela, I., and Harpaz, I., Further studies on the biology of an antiviral factor (AVF) from virus-infected plants and its association with the N-gene of *Nicotiana* species, *J. Gen. Virol.*, 35, 107, 1977.

248. Sela, I., Plant-virus interactions related to resistance and localization of virus infections, *Adv. Virus Res.*, 26, 201, 1982.

249. Sela, I., Interferon-like factor from virus-infected plants, *Perspect. Virol.*, 11, 129, 1981.

250. Sela, I., Antiviral factor from virus-infected plants, *Trends Biochem. Sci.*, 6, 31, 1981.

251. Orchansky, P., Rubinstein, M., and Sela, I., Human interferons protect plants from virus infection, *Proc. Natl. Acad. Sci. U.S.A.*, 79, 2278, 1982.

252. Ogarkov, V. I., Kaplan, I. B., Taliansky, M. E., and Atabekov, J. G., Reduction of potato viruses reproduction by human interferon (Russian), *Dokl. Akad. Nauk SSSR*, 276, 743, 1984.

253. Carter, W. A., Swartz, H., and Gillespie, D. H., Independent evolution of antiviral and growth-modulating activities of interferon, *J. Biol. Response Modifiers*, 4, 447, 1985.

254. Rosenberg, N., Reichman, M., Gera, A., and Sela, I., Antiviral activity of natural and recombinant human leukocyte interferon in tobacco protoplasts, *Virology*, 140, 173, 1985.

255. Vicente, M., De Fazio, G., Menezes, M. E., and Golgher, R. R., Inhibition of plant viruses by human gamma interferon, *J. Phytopathol.*, 119, 25, 1987.

256. Antoniw, J. F., White, R. F., and Carr, J. P., An examination of the effect of human interferons on the infection and multiplication of tobacco mosaic virus in tobacco, *Phytopathol., Z.*, 109, 367, 1984.

257. Huisman, M. J., Broxterman, H. J. G., Schellekens, H., and van Vloten-Doting, L., Human interferon does not protect cowpea protoplasts against infection with alfalfa mosaic virus, *Virology*, 143, 622, 1985.

258. Loech-Fries, L. S., Halk, E. L., Nelson, S. E., and Krahn, K. J., Human leucocyte interferon does not inhibit alfalfa mosaic virus in protoplast or tobacco tissue, *Virology*, 143, 626, 1985.

259. Kaplan, I. B., Taliansky, M. E., Malyshenko, S. I., Ogarkov, V. I., and Atabekov, J. G., Effect of human interferon on reproduction of plant and mycoviruses, *Arch. Phytopath. Pflanzenschutz*, 24, 3, 1988.

260. Edelbaum, O., Ilan, N., Grafi, G., Sher, N., Stram, Y., Novick, D., Tal, N., Sela, I., and Rubinstein, M., Purification and characterization of two antivirally active proteins from tobacco by monoclonal antibodies to human β-interferon, *Proc. Natl. Acad. Sci. U.S.A.*, 87, 588, 1990.

261. Edelbaum, O., Sher, N., Rubinstein, M., Novick, D., Tal, N., Moyer, M., Ward, E., Ryals, J., and Sela, I., Two antiviral proteins, gp35 and gp22, correspond to β-interferon-1,3-glucanase and an isoform of PR-5, *Plant Mol. Biol.*, 17, 171, 1991.

262. Livne, B., and Sela, I., unpublished data, 1995.

263. Samuel, C. E., Antiviral actions of interferon. Interferon-regulated cellular proteins and their surprisingly selective antiviral activities, *Virology*, 183, 1, 1991.

264. Devash, Y., Biggs, S., and Sela, I., Multiplication of tobacco mosaic virus in tobacco leaf disks is inhibited by (2′-5′) oligoadenylate, *Science*, 216, 1415, 1982.

265. Reichman, M., Devash, Y., Suhadolnik, R. J., and Sela, I., Human leucocyte interferon and the antiviral factor (AVF) from virus-infected plants stimulate plant tissue to produce nucleotides with antiviral activity, *Virology*, 128, 240, 1983.

266. Devash, Y., Gera, A., Willis, D. H., Reichman, M., Pfleiderer, W., Charubala, R., Sela, I., and Suhadolnik, R. J., 5′-Dephosphorylated 2′,5′-adenylate trimer and its analogs. Inhibition of tobacco mosaic virus replication in tobacco mosaic virus-infected leaf discs, protoplasts and intact tobacco plants, *J. Biol. Chem.*, 259, 3482, 1984.

267. Devash, Y., Reichman, M., Sela, I., Reichenbach, N. L., and Suhadolnik, R. J., Plant oligoadenylates: enzymatic synthesis, isolation and bilogical activities, *Biochemistry*, 24, 593, 1985.

268. Sher, N., Edelbaum, O., Barak, Z., Stram, Y., and Sela, I., Induction of an ATP-polymerizing enzyme in TMV-infected tobacco and its homology to human 2´ 5´A synthetase, *Virus Genes*, 4, 27, 1990.

269. Kulaeva, O. N., Fedina, A. B., Burkhanova, E. A., Karavaiko, N. N., Karpeisky, M. Y., Kaplan, I. B., Taliansky, M. E., and Atabekov, J. G., Biological activities of human interferon and 2´ 5´ oligoadenylates in plants, *Plant Mol. Biol.*, 20, 383, 1992.

270. Edelbaum, O., Stein, D., Holland, N., Gafni, Y., Livneh, O., Novick, D., Rubinstein, M., and Sela, I., Expression of active human interferon-b in transgenic plants, *J. Interferon Res.*, 12, 449, 1992.

271. Taliansky, M. E., personal communication, 1995.

272. Skryabin, K. G., personal communication, 1995.

273. Halperin, E., Edelbaum, O., and Sela, I., unpublished data.

274. Truve, E., Aaspollu, A., Honkanen, J., Puska, R., Mehto, M., Hassi, A., Teeri, T. H., Kelve, M., Seppanen, P., and Saarma, M., Transgenic potato plants expressing mammalian 2´ 5´ oligoadenylate synthetase are protected from potato virus X infection under field conditions, *Biotechnology*, 11, 1048, 1993.

275. Sela, I., Grafi, G., Sher, N., Edelbaum, O., Yagev, H., and Gerassi, E., Resistance systems related to the N gene and their comparison with interferon, in *Plant Resistance to Viruses*, Ciba Foundation Symposium 133, Evered, D. and Harnett, S., Eds., John Wiley & Sons, Chichester, U.K., 1987, 109.

276. Babosha, A. V., Trofimets, L. N., and Ladygina, M. E., Oligoadenylates and oligoadenylate synthetase of potato plants in protective reactions against virus pathogen (Russian), *Dokl. Akad. Nauk SSSR*, 313, 252, 1990.

277. Taylor, I. B., Biosystematics of tomato, in *The Tomato Crop*, Atterton, J. G. and Rudich, J., Eds., Chapman and Hall, London, 1986, 1.

278. Gebre-Selassie, K. and Marchoux, G., Identification and characterization of tobamovirus strains infecting L-resistant genotypes of peppers in France, *J. Phytopathol.*, 131, 275, 1991.

279. Cockerham, G., Genetical studies on resistance to potato viruses X and Y, *Heredity*, 25, 309, 1970.

280. Kohm, B. A., Goulden, M. G., Gilbert, J. E., Kavanagh, T. A., and Baulcombe, D. C., A potato virus X resistance gene mediates an induced, nonspecific resistance in protoplasts, *Plant Cell*, 5, 913, 1993.

281. Holmes, F. O., Inheritance of resistance to tobacco mosaic disease in tobacco, *Phytopathology*, 28, 553, 1938.

282. Padgett, H. S. and Beachy, R. N., Analysis of a tobacco mosaic virus strain capable of overcoming N gene mediated resistance, *Plant Cell*, 5, 577, 1993.

283. Culver, J. N. and Dawson, W. O., Tobacco mosaic virus elicitor coat protein genes produce a hypersensitive phenotype in transgenic *Nicotiana sylvestris* plants, *Mol. Plant Microbe Interact.*, 4, 458, 1991.

284. Tanksley, S. D., Young, N. D., Paterson, A. H., and Bonierbale, M. W., RFLP mapping in plant breeding: new tools for an old science, *Biotechnology*, 7, 257, 1989.

285. Dax, E., Livneh, O., Edelbaum, O., Kedar, N., Gavish, N., Karchi, H., Milo, J., Sela, I., and Rabinowitch, H. D., A random amplified polymerific DNA (RAPD) molecular marker for the Tm-2a gene in tomato, *Euphytica*, 74, 159, 1994.

286. Baker, B., Schell, J., Lorz, H., and Fedoroff, N., Transposition of the maize controlling element "Activator" in tobacco, *Proc. Natl. Acad. Sci. U.S.A.*, 83, 4844, 1986.

287. Whitham, S., Dinesh-Kumar, S. P., Choi, D., Hehl, R., Corr, C., and Baker, B., The product of the tobacco mosaic virus resistance gene N: similarity to toll and the interleukin-1 receptor, *Cell*, 78, 1101, 1994.

288. Carr, J. P. and Klessig, D. F., The pathogenesis-related proteins of plants, in *Genetic Engineering, Principles and Methods*, Vol. 11, Setlow, J. K., Ed., Plenum Press, New York, 1979, 65.

289. Cutt, J. R., Harpster, M. H., Dixon, D. C., Carr, J. P., Dunsmuir, P., and Klessig, D. F., Disease response to tobacco mosaic virus in transgenic tobacco plants that constitutively express the pathogenesis-related PR1b gene, *Virology*, 173, 88, 1989.

290. Linthorst, H. J. M., Meuwissen, R. L. J., Kauffmann, S., and Bol, J., Constitutive expression of pathogenesis-related proteins PR-1, GRP, and PR-S in tobacco has no effect on virus infection, *Plant Cell*, 1, 286, 1989.

291. Becker, F., Buschfeld, E., Schell, J., and Bachmair, A., Altered response to viral infection by tobacco plants perturbed in ubiquitin system, *Plant J.*, 3, 875, 1993.

292. Bawden, F. C., Inhibitors and plant viruses, *Adv. Virus Res.*, 2, 31, 1954.

293. Babieri, L., Aron, G. M., Irvin, J. D., and Stripe, F., Purification and partial characterization of another form of the antiviral protein from the seeds of *Phytolacca americana* L. (pokewood), *Biochem. J.*, 203, 55, 1982.

294. Zarling, J. M., Moran, P. A., Haffar, O., Sias, J., Richman, D. D., Spina, C. A., Mayers, D. A., Kuebelbeck, V., Ledbetter, J. A., and Uckun, F. M., Inhibition of HIV replication by pokeweed antiviral protein targeted to CD4+ cells by monoclonal antibodies, *Nature*, 347, 92, 1990.

295. Ready, M. P., Brown, D .T., and Robertus, J. D., Extracellular localization of pokeweed antiviral protein, *Proc. Natl. Acad. Sci. U.S.A.*, 83, 5053, 1986.

296. Lodge, J. K., Kaniewski, W. K., and Tumer, N. E., Broad-spectrum virus resistance in transgenic plants expressing pokeweed antiviral protein, *Proc. Natl. Acad. Sci. U.S.A.*, 90, 7089, 1993.

297. De Zoeten, G. A., Risk assessment: do we let history repeat itself?, *Phytopathology*, 81, 585, 1991.

298. Rochow, W. F., Barely yellow dwarf virus: phenotypic mixing and vector specificity, *Science*, 167, 875, 1970.

299. Osbourn, J. K., Sarkar, S., and Wilson, T. M. A., Complementation of coat protein-defective TMV mutants in transgenic tobacco plants expressing TMV coat protein, *Virology*, 179, 921, 1990.

300. Atreya, C. D., Raccah, B., and Pirone, T. P., A point mutation in the coat protein abolishes aphid transmissibility of a potyvirus, *Virology*, 178, 161, 1990.

301. Gal-On, A., Antignus, Y., Rosner, A., and Raccah, B., A zucchini yellow mosaic virus coat protein gene mutation restores aphid transmissibility but has no effect on multiplication, *J. Gen. Virol.*, 73, 2183, 1992.

302. Bourdin, D. and Lecoq, H., Evidence that heteroencapsidation between two potyviruses is involved in aphid transmission of a non-aphid transmissible isolate from mixed infections, *Phytopathology*, 81, 1459, 1991.

303. Candelier-Harvey, P. and Hull, R., Cucumber mosaic virus genome is encapsidated in alfalfa mosaic virus coat protein expressed in transgenic tobacco plants, *Transgenic Res.*, 2, 277, 1993.

304. Smith, D. B. and Inglis, S. C., The mutation rate and variability of eucaryotic viruses: an analytical review, *J. Gen. Virol.*, 68, 2729, 1987.

305. King, A. M. Q., Genetic recombination in positive strand RNA viruses, in *RNA Genetics*, Vol. 2, Domingo, E., Holland, J. J., and Ahlquist, P., Eds., CRC Press, Boca Raton, FL, 1988, 149.

306. Shirako, Y. and Brakke, M. K., Spontaneous deletion mutation of soil-born wheat mosaic virus II, *J. Gen. Virol.*, 65, 855, 1984.

307. Rujarski, J. J. and Kaesberg, P., Genetic recombination between RNA components of a multiple plant virus, *Nature*, 321, 528, 1986.

308. Robinson, D. J., Hamilton, W. D. O., Harrison, B. D., and Baulcombe, D. C., Two anomalous tobravirus isolates: evidence for RNA recombination in nature, *J. Gen. Virol.*, 68, 2551, 1987.

309. Hillman, B. I., Carrington, J. C., and Morris, T. J., A defective interfering RNA that contains a mosaic of plant virus genome, *Cell*, 51, 427, 1987.

310. Bouzoubaa, S., Niesbach-Klosgen, U., Jupin, I., Guilley, H., Richards, K., and Jonard, G., Shortened forms of beet necrotic yellow vein virus RNA-3 and -4: internal deletions and subgenomic RNA, *J. Gen. Virol.*, 72, 259, 1991.

311. Allison, R., Thompson, C., and Ahlquist, P., Regeneration of a functional RNA virus genome by recombination between deletion mutants and requirement of cowpea chlorotic mottle virus 3a and coat genes from systemic infection, *Proc. Natl. Acad. Sci. U.S.A.*, 87, 1820, 1990.

312. Bujarski, J. J. and Dzianott, A. M., Generation and analysis of nonhomologous RNA RNA recombination in brome mosaic virus: sequence complimentarities at crossover sites, *J. Gen. Virol.*, 65, 4153, 1991.

313. Bujarski, J. J., Nagy, P. D., and Flasinski, S., Molecular studies of genetic RNA RNA recombination in brome mosaic virus, *Adv. Virus Res.*, 43, 275, 1994.

314. Greene, A. E. and Allison, R. F., Recombination between viral RNA and transgenic plant transcripts, *Science*, 263, 1423, 1994.

315. Falk, B. W. and Bruening, G., Will transgenic crops generate new viruses and new diseases?, *Science*, 263, 1395, 1994.

Chapter 7

GENETICALLY ENGINEERED FUNGI IN AGRICULTURE

Nancy L. Brooker and William Bruckart

TABLE OF CONTENTS

1   INTRODUCTION

1.1   Potential for Fungal Manipulation

Fungi are used in a number of applications, including food production (e.g., cheese, mushrooms, koji), enzyme production (e.g., glucoamylase, cellulase, alpha-amylase, proteases, lipases, etc.), and production of primary and secondary metabolites (e.g., amino acids, organic acids, antibiotics, and gibberellins). Recently, fungi have been considered for commercial use in the biological control of weeds, insects, and plant disease pests, and there is the potential for use of fungi in bioremediation.[1]

Current molecular genetic techniques have been particularly useful in modifying fungi. Attributes that make fungi excellent candidates for genetic modification include relatively small eukaryotic genomes, high levels of inherent variability with which to select for commercially important traits, ease of monoculturing and biomass generation, and the ability to manipulate specific metabolically important traits *in vitro*.

1.1.1   *Physiological and Morphological Traits*

Fungi possess traits not found in other plant or microbial systems.[2] Among these traits are capability for direct penetration of plant and animal hosts, production of toxins and enzymes unique to pathogenesis of plants and animals, and the ability to grow rapidly and produce large numbers of propagative units.[3,4]

1.1.2   *Suitability for Genetic Modification and Selection*

Many classical genetic studies have utilized *Neurospora* and *Aspergillus* spp. in model systems to learn about metabolism, genetics, and physiology. The procedures for studying *Neurospora* and *Aspergillus* spp. are not necessarily applicable to the study of pathogenic fungi. Methods for protoplast generation and transformations have required modification for studying fungi other than *Aspergillus* and *Neurospora* spp.[5]

There are several ways to obtain desirable characteristics in plant pathogenic fungi, including screening, sexual reproduction and parasexual processes, protoplast fusion, directed mutagenesis, and recombinant DNA engineering.

1.2   Plant Pathogenic Fungi

Fungi are important agriculturally as plant pathogens, both as crop pests and as agents for biological control of plant diseases and weeds. In this review we cite examples of research on fungi intended for biocontrol, but the processes described for improvement and alteration of fungi in biocontrol are similar to those for other applications.[5] Examples are cited to illustrate both accomplishments and the potential of genetic engineering with fungi; they are not intended to reflect a thorough review of this subject.

There are two basic tactics used to deploy plant pathogens for weed control, the classical (inoculative) and the bioherbicide (inundative mycoherbicide) approaches.[6-8] Currently, three registered mycoherbicide

products are available for commercial use. They are Collego™*
(*Colletotrichum gloeosporioides* f.sp. *aeschynomene*), for the control of northern jointvetch, *Aeschynomene virginica*; DeVine™** (*Phytophthora palmivora*),
for the control of stranglervine, *Morrenia odorata*; and Biomal™***
(*Colletotrichum gloeosporioides* f.sp. *malvae*), for the control of round-leafed
mallow, *Malva pusilla*.[9-13]

These fungal products have the essential characteristics for an effective biocontrol agent: aggressiveness towards an important weed, host
specificity, spore production, consistency and performance in the field,
long shelf life, ease of application using standard farming equipment, and
a suitable commercial market. These developed and other potential
mycoherbicides may be "improved", however, using the following strategies. Specific areas of improvement that have been considered are a)
pathogenicity and virulence, b) dissemination, c) competitive ability, d)
detection and identification, and e) safety.[14]

a) **Pathogenicity and Virulence.** Pathogenicity and virulence can be altered in either a qualitative or a quantitative manner. Specific alterations include increased rate of infection, shorter disease cycles, increased production of infectious propagules, and host range alterations.

b) **Dissemination.** Most fungi have complex life cycles that ensure both
   dissemination and survival. Life state components include multiple
   disease cycles involving asexually produced spores, sexual reproduction, and production of spores that are resistant to adverse environmental conditions.

c) **Competitive Abilities.** Individual strains selected from within a fungal
   population may possess greater competitive abilities than the population as a whole. The magnitude of intraspecies variation in competitiveness presents ideal opportunities for strain improvement.

d) **Detection and Identification.** Mutation of strains by irradiation or
   chemicals and screening for spontaneous changes offers methods of
   acquiring unique, readily identifiable strains for research. Marked strains
   that contain drug resistance or other unique mutations enable rapid
   identification and monitoring of fungi in experimental situations.

e) **Safety.** The use of nutritional auxotrophs and the potential of host- or
   environment-dependent expression of virulence genes represent a new
   direction in improving the safety of biocontrol fungi.

To improve these traits, it is necessary to understand the processes
behind them. Introducing new genes into an organism suggests that there

* Registered Trademark of Ecogen, Inc., Langhorne, Pennsylvania, United States.
** Registered Trademark of Abbott Laboratories, Chicago, Illinois, United States.
*** Registered Trademark of Philom Bios, Inc., Saskatoon, Saskatchewan, Canada.

is understanding of disease processes, virulence, and host specificity. Actually, only a few genes have been isolated that play known roles in disease development. Little is known about the relationship between genotype and specific traits such as virulence.[14] Utilizing the techniques currently available for modification of phytopathogenic fungi in biological control may lead to new principles and concepts of the disease process.

2   GENETIC ENGINEERING OF FUNGI IN AGRICULTURE

There is no practical field experience with engineered fungi in plant pathology. Most effort has been directed towards engineering fungi for research on plant disease processes; however, this information has direct relevance to biocontrol concepts. The two genetic engineering approaches most commonly pursued in attempts to improve biocontrol fungi have been insertion of genes and deletion of genes. Genes are inserted that code for a specific phenotypic trait, such as improved pathogenicity, virulence, host range, and fungicide or toxin resistance. The benefits of gene insertion are a specific and verifiable alteration of the genome and potentially greater stability. This stability is the result of reduced numbers of nonspecific effects on the genome. This approach, however, is limited by the identification and cloning of only a few genes that code for these characteristics.

Gene deletion generally utilizes mutation of an effective nonspecific pathogen and the restriction of its host range. Mutations can be generated by chemical or physical means or by using recombinant molecular genetics methods. Benefits of gene deletion include the number of pathogens suitable for manipulation and the potential for success. Deletion mutants can also be monitored easily by phenotypic or biochemical procedures; however, they can be unstable if the deletion is not specific or if the potential for reversion is high.

2.1   Environmental Release and Regulations

No field tests of genetically engineered fungi have been conducted under permit from the Animal and Plant Health Inspection Service (APHIS), which has regulated environmental releases of certain genetically modified organisms since 1987 under the authority of the Federal Plant Pest Act and the Plant Quarantine Act. In addition, the Environmental Protection Agency (EPA) has not received any field test submissions under the Federal Insecticide, Fungicide, and Rodenticide Act (FIFRA) or the Toxic Substances Control Act (TSCA). Pesticidal substances are regulated under FIFRA, and biocontrol agents are considered pesticides. TSCA regulates microorganisms that are not pesticides and are not intended for food or drug use. Modified fungi used for bioremediation purposes would be regulated under TSCA.[15,16]

2.2   Research

Current research emphasizes the study of plant pathogenic fungi, rather than development of biocontrol agents. The focus of these studies generally concerns the basis of host-pathogen interactions, i.e., genes specifically involved with pathogenicity and virulence. The scope of this chapter is to highlight research findings which have relevance in biocontrol efficacy. Research is heavily biased towards host-pathogen interaction studies.

2.2.1   *Pathogenesis — Enzymes*

2.2.1.1   *Phytoalexin Degradation*

VanEtten et al. have shown that the pea root rot fungus, *Nectria haematococca*, is resistant to microbe-induced low-molecular-weight defense compounds (phytoalexins) in peas.[17] The major phytoalexin in peas, pisatin, can be detoxified through demethylation by pisatin demethylase (PDA), an enzyme produced by *N. haematococca*.[17] The PDA gene was used to transform a non-PDA-producing, nonpathogenic variant strain of *N. haematococca*. Three transformants were able to demethylate pisatin *in vitro*, and two of the three were pathogenic on pea. Theses results suggested a pathogenicity role for PDA.[18-20] However, results of gene segregation and gene disruption studies of the PDA gene contradict this pathogenicity role.[21,22] PDA is now proposed to have a subtle role in pathogenicity. Data indicate that there may be some other genes linked to the PDA genes on their dispensable chromosomes that determine pathogenicity.[22]

This study highlights the complexity of host-pathogen interactions and the need for a more detailed biochemical understanding of plant disease resistance. However, the use of molecular biology techniques definitively elucidated the role of PDA in pathogenicity. By understanding the various chemical defenses within plants, it may be possible to develop better and more efficient biocontrol agents. Such information also will lead to a better understanding of the potential of biocontrol agents and strategies for their use.

2.2.1.2   *Cell-Degrading Enzymes*

Kolattukudy has emphasized the importance of physical barriers as plant defense mechanisms.[23] Cutinase is a fungal enzyme which degrades the plant cutin layer. Cutin is made up of fatty acid polymers and represents a major structural component of the epidermis of plant leaves and fruits. The capacity of many fungi to penetrate the plant cutin barrier is sufficient for infection to proceed.[23]

Colletotrichum gloeosporioides must produce cutinase in order to infect the papaya plant.[24] By comparison, *Mycosphaerella* spp. normally require wounds to breach the cuticle before infection occurs. Exogenous cutinase in the presence of *Mycosphaerella* spp. increases infection efficiency, and transformants of *Mycosphaerella* with the cutinase gene from *Fusarium oxysporum* produced cutinase and infected papaya fruit that had not been wounded. Inhibition of infection by the *Mycosphaerella* transformants occurred in the presence of a cutinase antibody.[25] However, cutinase gene disruption experiments of *N. haematococca* and *Magnaporthe grisea* showed that cutinase is not essential for the infection of pea or rice, respectively.[26-29]

Other cell-wall-degrading enzymes have been studied to determine their role in pathogenesis.[30] A mutant of *Cochliobolus carbonum*, a pathogen of maize, lacking in endopolygalacturonase (EPG) activity was created by insertionally inactivating the enzyme with an internal homologous fragment of the gene.[31] Compared with wild-type isolates, the mutant without EPG activity was equally virulent on maize, suggesting a more complex and subtle role for this cell-degrading enzyme.

Genetic engineering has allowed for detailed studies of PDA, cutinase, and EPG in pathogenicity roles. New or improved biocontrol agents may be developed by engineering them with genes coding for enzymes which affect pathogenicity or virulence.

2.2.2   Pathogenesis — Toxins

Transformants with genes for phytotoxin production may have biocontrol capabilities. Genes that code for production of phaseolotoxin, tabtoxin, bialaphos, and other toxins, such as those from *Alternaria* and *Helminthosporium* may affect host range and virulence.[32-38] Bacteria-produced toxins are coded by genes in clusters or operons within the prokaryote genome. Transformation and expression of prokaryote gene clusters in eukaryotic fungi represents a new and challenging area of research.

2.2.3   Pathogenesis — Fitness

The potential to improve fitness, the individual's contribution to the gene pool of the next generation, has been clearly identified in studies with *Trichoderma harzianum* and *N. haematococca*. Strains of *T. harzianum* created by protoplast fusion were screened on the basis of nutritional, morphological, isoenzymatic, and growth characteristics. Isolates were first screened in culture to remove variability, and then single-spore progeny were tested for similarity. One strain was identified as being superior to either parent for seed protection and rhizosphere competence.[39]

The pea pathogen discussed earlier, *N. haematococca*, is not normally a pathogen of tomato. Green tomato fruits are not susceptible, but ripe fruits become necrotic when inoculated with *N. haematococca*.[40,41] Unripened (green) tomato fruits have high levels of tomatine, an inhibitory steroidal glycoalkaloid, compared with ripened fruits. A tomatine-resistant mutant of *N. haematococca* infected green tomato fruit. Genetic studies indicated complete linkage and monogenic inheritance of tomatine resistance in *N. haematococca* related to the increased aggressiveness on green tomato fruits.[40,41] This example illustrates the usefulness of genetic analysis to confirm the genetic characteristics of this change. In addition, knowledge about probable mechanisms for resistance enable scientists to design an experiment to overcome plant resistance.[14]

Genetic transformation and fitness have been studied in the plant pathogen *Cochliobolus heterostrophus*, the causal agent of Southern corn leaf blight.[42] Two different enzymatic markers, hygromycin B phosphotransferase, which detoxifies hygromycin B, and acetamidase, which allows for growth on media containing acetamide as the sole nitrogen source, were used in four different plasmid constructs. Fitness of transformed fungi was compared to that of the isogenic wild-type progenitor in competitive pathogenicity tests on maize plants. Transformants were found to be less fit than the wild type in 92% of the tests. However, further tests indicated that the protocol used to make the protoplasts generated phenotypic instability and progeny that were less fit. Factors which had no impact on fitness included size and copy number of the transforming plasmid, site of plasmid integration into chromosomal DNA, temperature at which the plant assays were performed, proportion of initial inoculum composed of transformant conidia, and recipient fungal strain.[42] This study presents a very interesting point: that preparatory processes (e.g., protoplasting) may create instability and reduce fitness in a fungus. These results lead to questions about the importance of nuclear number, nuclear disruption, and the role of nuclear regulation in fungal fitness.

2.2.4   Genetically Engineered Strain Modifications

2.2.4.1   Fungicide Resistance

Unique strategies involving the genetic engineering of biocontrol fungi are currently being used to enhance cultural characteristics. *Gliocladium virens*, used for biocontrol of damping-off diseases in zinnia, cotton, and cabbage in the greenhouse, is not effective where the fungicide Benomyl™* is used. Transformants of *G. virens* with a Benomyl resistance gene can be used in the presence of Benomyl.[43] Integration of such improved biocontrol agents with other pest control strategies facilitates pest management by coapplication of fungicide and biocontrol agent. Results of this study demonstrate how genetic engineering may be used to specifically improve a biocontrol characteristic, thus improving its efficacy. Additionally, this approach enables lower chemical usage for fungal disease control.

---

* Registered Trademark of E. I. du Pont de Nemours and Company, Inc., Wilmington, Delaware.

Colletotrichum gloesporioides f.sp. *aeschynomene,* a commercially available biocontrol agent mentioned previously, and *C. graminicola,* causal agent of corn anthracnose, have also been engineered with Benomyl resistance.[44,45]

2.2.4.2   Herbicide Resistance

Genetic engineering has been used to transform species of *Colletotrichum* with the bialaphos acetyl transferase (bar gene) from *Streptomyces hygroscopicus.* Bialaphos is metabolized by plants, fungi, and other eukaryotes to produce a toxic analogue of glutamic acid which is lethal. The bar gene codes for acetylation and inactivation of the toxic analogue. Transformants with the bar gene are bialaphos resistant. *Colletotrichum gloeosporioides* f.sp. *aeschynomene* and *C. coccodes* transformed with the bar gene display bialaphos resistance as high as 100 times that of the wild type.[46] Coapplication of the transformed *Colletotrichum* species with sublethal doses of the herbicide bialaphos is being studied for its potential in weed control and to assess the effect of bialaphos on *Colletotrichum* virulence and host range.[47] This coapplication approach may enable reduction in the rates of both herbicide and fungal inoculum concentration for weed control.

2.2.4.3   Host-Dependent Gene Expression

In addition to auxotrophic mutants, a proposal has been made to use host-inducible promoter segments to control expression of genes engineered in fungi.[48] This approach may allow for a more intimate expression of virulence-related genes and may be incorporated for reasons of safety and management of biocontrol agents. Without the appropriate host, the biocontrol agent will not be "activated".

2.2.4.4   Suicide Vectors

Conversely, genes inactivated by the presence of a specific substrate, or suicide vectors, represent another approach to managing genetically engineered fungi. Suicide genes are activated in the presence of a specific substrate, and the organism self-destructs.[49,50]

2.2.4.5   Spatial Gene Isolation

Another approach to containment and safety is to split gene expression among several vectors or genomes, and gene expression requires all vectors or genomes to be present.[51] By spatially separating the genes needed for expression, there is a temporal mode of regulating gene expression.

2.2.4.6   Disruption of the Mating-Type Loci

Disruption of the mating-type genes within biocontrol fungi with known sexual stages might also be used to limit the spread of engineered genes in the environment. Theoretically, the spread of foreign genes in natural systems could be prevented by requiring that the genetically engineered strains be altered in their mating-type genes.[52]

2.2.5   Nonengineered Strategies for Modification of Fungi

2.2.5.1   Auxotrophs

Phenotypic and biochemical mutants of *Sclerotinia sclerotiorum* were obtained by ultraviolet irradiation. One phenotypic mutant lacked the ability to produce sclerotia, structures required both for overwintering and sexual reproduction. This strain did not survive adverse weather conditions or produce ascospores in a limited field trial.[53] Cytosine auxotrophic mutants of *S. sclerotiorum*, which require an external source of the pyrimidine to cause infection, also were studied in the field.[54] This strain produces apothecia and ascospores which also need cytosine to survive. Since exogenous cytosine does not occur naturally outside of the field site, the organism did not disperse or persist. The use of biochemical and developmental fungal mutants is a novel approach to containment and marking of mutant fungi. Although these isolates were derived by directed mutagenesis, these strategies may be developed and further enhanced through genetic engineering procedures.

2.2.5.2   Spontaneous Genetic Mutations

Another approach for tracking fungal populations is through the use of Nit mutants.[55-57] Nit mutants occur spontaneously on a nitrate analogue, chlorate. The mutation is in the nitrate utilization pathway, in the nitrate reductase structural genes. Nit mutants of *Fusarium oxysporum* have been used to chart vegetative compatibility groups, population dynamics, and levels of relatedness, thereby permitting the study of populations in the field.[56] No differences were found when Nit mutants of *C. gloeosporioides* f.sp. *aeschynomene* were compared with the parental strains for virulence.[57] However, great phylogenetic diversity was confirmed within the genus *Colletotrichum*. Nit mutants have demonstrated potential use in fungal studies including phylogenetics, population dynamics, and tracking modified strains in natural systems.

2.2.6   Summary

These listed approaches are hypothetical for fungi since many of the recombinant techniques, regulatory mechanisms, and specific genes required for such manipulations have not been isolated in phytopathogenic fungi. However, these approaches offer unique modes of temporally and spatially controlling genetically engineered microorganisms.

3 CONCERNS FOR ENGINEERED FUNGI

All fungi of agricultural importance are considered in various state and federal regulations. Concerns and issues surrounding the use of engineered fungi are the same as for other engineered organisms. The current approach to regulation of microbials for agricultural use focuses primarily on the final product rather than the process used to generate the new strain, i.e., product vs. process. Although a number of approaches can be used to achieve a desirable characteristic, as discussed earlier in this chapter, the end function is the same. Except for gene transfer procedures, alterations result either from modifications of the target organism genome or from selection from natural variation within the population. These are generally considered low risk, since the processes used to develop such traits are more or less natural. Some of the goals of genetic engineering, such as the deletion of genes, also mimic the natural process. Applications for some altered organisms, including those which are genetically engineered, occur in highly contained systems such as fermentation for a specific biochemical product. These processes and applications should be considered low risk.

Fear of genes not naturally found in the target organism and from the lack of knowledge about how these genes will function in natural systems must be addressed before genetically engineered fungi may be used in proposed field applications.[58-62] There is a potential for unexpected damage to or displacement of nontarget organisms resulting form untested use of genes or their products in new organisms in nature. Furthermore, certain applications of these organisms may require large doses, similar to the current practice of using endemic fungi as bioherbicides. The little knowledge we have about variability and stability of fungi in natural populations exacerbates the absence of practical field experience with fungi altered by genetic engineering or other means. Questions of risk associated with genetically engineered fungi and the proposed use of the modified organism relate directly to the current understanding of fungal genetics and ecology.[60-63]

However, risk can be measured and the necessary judgments can be made for many engineered fungi. These decisions would be based on the knowledge of the gene used and how it functions in its source. Comparisons can be made on how a gene functions in its new host, and studies on stability of the integration can be conducted. Research also can be conducted to identify and quantify nontarget effects and the potential for gene transfer between species related and unrelated to the target organism. In addition, there may be information available about the gene in nature; it may be common or unique. Other considerations include

1) Is the altered gene immortal, or is the gene constructed in such a way as to assure immutability?[64] Such constructions are often associated with genes essential for existence, where the loss of all copies of the gene results in lethality to the organism.

2) Is the gene eliminated from the population via stabilizing selection?[60,65] Unless the gene product confers an advantage or is at least neutral to the organism, it is usually eliminated in nature.

3) Is the manipulated gene promiscuous; can the gene be transferred interspecifically or intraspecifically after release into the ecosystem?[66-72]

4) Are multiple copies of a given gene, or multiple rearrangements, integrated into the organism to ensure phenotypic and genotypic stability?[67]

Since the best risk assessment uses the statistical analysis of historical data, and these data do not exist for genetically engineered fungi, it is impossible to accurately predict the consequences of such an action as introduction of an engineered fungus into nature.[73] It has been proposed that resources would be better spent on measuring the potential benefits of this technology and to develop systems that reduce risks to acceptable levels.[74] This would be more constructive than to continue with futile attempts to assess risk with insufficient information.[2] The needed experience will occur only if field tests are permitted with engineered organisms of low risk. Steps that will lead to a better understanding of genetically engineered organisms in the field have been proposed by Levin and Harwell and include[75]

1) Use a case-by-case assessment, before release, of the fate and transport of the genetic information and of possible effects of introduced genes.

2) Include as part of the release protocol procedures to monitor fate, transport, and effects after release.

3) Use a plan for containment within predefined spatial and temporal limits, where practical.

4) Utilize a contingency plan for mitigation in case of undesirable side effects, where needed.

Over the last 8 years, 11 reviews have addressed the subject of engineered fungi and their fitness in the environment.[1,2,6-8,14,55,61,74,76,77] However, there have been no field tests of such organisms to date. The numerous reviews on this subject combined with the absence of practical field experience highlight the interest in, and frustration with, this research. New approaches to risk assessment and potential benefits may lead to practical use of genetically engineered fungi. Also, increased public awareness and support for this approach will speed progress towards practical utilization of these organisms. However, until there is more knowledge of fungal ecology and population genetics, it will be necessary to review proposals on a case-by-case basis.

There is considerable potential to develop much of the needed information using safe genetically engineered organisms in field studies. As the available data increases, more realistic judgments will be possible for risk assessment of engineered organisms.

REFERENCES

1. Rambosek, J. and Leach, J., Recombinant DNA in filamentous fungi: progress and prospects, *CRC Crit. Rev. Biotechnol.*, 6:357, 1987.
2. Miller, R. V. and Sands, D. C., Fitness of genetically altered fungi, in *The Fungal Community*, Carroll, G. C. and Wicklow, D. T., Eds., Marcel Dekker, New York, 1992, chap. 6.
3. Alexopoulos, C. J. and Mims, C. W., *Introductory Mycology*, John Wiley & Sons, New York, 1979.
4. Agrios, G. N., *Plant Pathology*, Academic Press, New York, 1978.
5. Fincham, J. R. S., Transformation in fungi, *Microbiol. Rev.*, 53:148–170, 1989.
6. TeBeest, D. O., Yang, X. B., and Cisar, C. R., The status of biological control of weeds with fungal pathogens, *Annu. Rev. Phytopathol.*, 30:637, 1992.
7. TeBeest, D. O., Biological control of weeds: potential for genetically modified strains, in *Advanced Engineered Pesticides*, Kim, L., Ed., Marcel Dekker, New York, 1993.
8. Charudattan, R., The use of natural and genetically altered strains of pathogens for weed control, in *Biological Control in Agricultural IPM Systems*, Hoy, M. A. and Herzog, D. C., Eds., Academic Press, New York, 1985, p. 347.
9. Templeton, G. E., Mycoherbicides — achievements, developments and prospects, in *Proceedings of the Eighth Australian Weeds Conference*, Weed Society of New South Wales, Sydney, Australia, 1987, p. 489.
10. Templeton, G. E., TeBeest, D. O., and Smith, R. J., Jr., Biological weed control in rice with a strain of *Colletotrichum gloeosporioides* (Penz.) Sacc. used as a mycoherbicide, *Crop Prot.*, 3:409, 1984.
11. Ridings, W. H., Biological control of stranglervine in citrus — a researcher's view, *Weed Sci.*, 34:31, 1986.
12. Kenney, D. S., Devine — the way it was developed — an industrialist's view, *Weed Sci.*, 34:15, 1986.
13. Mortensen, K., The potential of an endemic fungus, *Colletotrichum gloeosporioides* f. sp. *malvae*, for biological control of round-leafed mallow (*Malvae pusilla*) and velvetleaf (*Abutilon theophrasti*), *Weed Sci.*, 36:473, 1988.
14. Kistler, H. C., Genetic manipulation of plant pathogenic fungi, in *Microbial Control of Weeds*, TeBeest, D. O., Ed., Chapman and Hall, New York, 1991, chap. 9.
15. Plant pests; introduction of genetically engineered organisms or products; final rule, *Fed. Regist.*, 52:115, 1987.
16. User's guide for introducing genetically engineered plants and microorganisms, *Fed. Regist.*, 50:184, 1985.
17. VanEtten, H. D., Mathews, D. E., and Mackintosh, S. F., Adaptation of pathogenic fungi to toxic chemical barriers in plants: the pisatin demethylase of *Nectria haematococca* as an example, in *Molecular Strategies for Crop Protection*, Alan R. Liss, New York, 1987, p. 59.
18. Weltring, K. M., Turgeon, B. G., Yoder, O. C., and VanEtten, H. D., Cloning a phytoalexin gene detoxification gene from the plant pathogenic fungus *Nectria haematococca* by expression in *Aspergillus nidulans*, *Gene*, 68:335, 1988.
19. Miao, V. P., Covert, S. F., and VanEtten, H. D., A fungal gene for antibiotic resistance on a dispensable ("B") chromosome, *Science*, 254:1773, 1991.
20. Ciuffetti, L. M., Weltring, K. M., Turgeon, B. G., Yoder, O. C., and VanEtten, H. D., Transformation of *Nectria haematococca* with a gene for pisatin demethylating activity and the role of pisatin detoxification in virulence, *J. Cell. Biochem. Suppl.*, 12C:278, 1988.
21. Funnell-Baerg, D. L., Mathews, P. S., and VanEtten, H. D., Unusual segregation of pathogenicity genes among laboratory crosses with a *Nectria haematococca* strain bred for virulence on pea, in Program of Abstracts of the 17th Fungal Genetics Meeting, Asilomar, CA, March 23 to 28, 1993.
22. VanEtten, H. D., personal communication, 1993.

23. Kolattukudy, P. E., Enzymatic penetration of the plant cuticle by fungal pathogens, *Annu. Rev. Phytopathol.*, 23:223, 1985.

24. Dickman, M. B. and Patil, S. S., Cutinase deficient mutants of *Colletotrichum gloeosporioides* are nonpathogenic to papaya fruit, *Physiol. Mol. Plant Pathol.*, 28:235, 1986.

25. Dickman, M. B., Podila, G. K., and Kolattukudy, P. E., Insertion of cutinase gene into a wound pathogen enables it to infect intact host, *Nature*, 342:446, 1989.

26. Stahl, D. J., Hannemann, F., Hoffmann, C., and Schafer, W., The role of cutinase during infection of pea by *Nectria haematococca*, in Program of Abstracts of 17th Fungal Genetics Meeting, Asilomar, CA, March 23 to 28, 1993.

27. Stahl, D. J. and Schafer, W., Cutinase is not required for fungal pathogenicity on pea, *Plant Cell*, 4:621, 1992.

28. Sweigard, J. A., Chumley, F. G., and Valent, B., Cloning and analysis of CUT1, a cutinase gene from *Magnaporthe grisea*, *Mol. Gen. Genet.*, 232:183, 1992.

29. Sweigard, J. A., Chumley, F. G., and Valent, B., Disruption of a *Magnaporthe grisea* cutinase gene, *Mol. Gen. Genet.*, 232:183, 1992.

30. Hahn, M. G., Bucheli, P., Cervone, F., et al., Roles of cell wall constituents in plant-pathogen interactions, in *Plant-Microbe Interactions*, Vol. 3, Kosuge, T. and Nester, E. W., Eds., McGraw-Hill, New York, 1989, p. 131.

31. Scott-Craig, J. S., Panacione, D. G., and Cervone, F., Endopolygalacturonase is not required for pathogenicity of *Cochliobolus carbonum* on maize, *Plant Cell*, 2:1191, 1990.

32. Peet, R. C., Lindgren, P. B., and Willis, D. K., Identification and cloning of genes involved in phaseolotoxin production by *Pseudomonas syringae* pv. *phaseolicola*, *J. Bacteriol.*, 166:1096, 1986.

33. Quigley, N. B., Lane, D., and Bergquist, P. L., Genes for phaseolotoxin synthesis are located on the chromosome of *Pseudomonas syringae* pv. *phaseolicola*, *Curr. Microbiol.*, 12:295, 1985.

34. Durbin, R. D. and Uchytil, T. F., The role of zinc in regulating tabtoxin production, *Experientia*, 41:136, 1985.

35. Kinscherf, T. G., Coleman, R. H., and Barta, T. M., Cloning and characterization of the tabtoxin biosynthesis region from *Pseudomonas syringae*, *J. Bacteriol.*, 173:4124, 1991.

36. Leason, M., Cunliffe, D., Parkin, D., et al., Inhibition of pea leaf glutamine synthetase by methionine suphoximine, phosphinothricin and other glutamate analogs, *Phytochemistry*, 21:855, 1982.

37. Ridley, S. M. and McNally, S. F., Effects of phosphinothricin on the isoenzymes of glutamine synthetase isolated from plant species which exhibit varying degrees of susceptibility to the herbicide, *Plant Sci.*, 39:31, 1986.

38. Murakami, T., Anzai, H., Imai, S., Satoh, A., Nagaoka, K., and Thompson, C. J., The bialaphos biosynthetic genes of *Streptomyces hygroscopicus*: molecular cloning and characterization of the gene cluster, *Mol. Gen. Genet.*, 205:42, 1986.

39. Harman, G. E. and Stasz, T. E., Protoplast fusion for the production of superior biocontrol fungi, in *Microbial Control of Weeds*, TeBeest, D. O., Ed., Chapman and Hall, New York, 1991, chap. 10.

40. Defago, G. and Kern, H., Induction of *Fusarium solani* mutants insensitive to tomatine, their pathogenicity and aggressiveness to tomato fruits and pea plants, *Physiol. Plant Pathol.*, 22:29, 1983.

41. Defago, G., Kern, H., and Sedlar, L., Genetic analysis of tomatine insensitivity, sterol content and pathogenicity for green tomato fruits in mutants of *Fusarium solani*, *Physiol. Plant Pathol.*, 22:39, 1983.

42. Keller, N. P., Bergstrom, G. C., and Yoder, O. C., Effects of genetic transformation on fitness of *Cochliobolus heterostrophus*, *Phytopathology*, 80:1166, 1990.

43. Ossanna, N. and Mischke, S., Genetic transformation of the biocontrol fungus *Gliocladium virens* to Benomyl resistance, *Appl. Environ. Microbiol.*, 56:3052, 1990.

44. Armstrong, J. K. and Harris, D. L., Biased DNA integration in *Colletotrichum gloeosporioides* f.sp. *aeschynomene* transformants with Benomyl resistance, *Phytopathology*, 83:328, 1993.

45. Panaccione, D. G., McKiernan, M., and Hanau, R. M., *Colletotrichum graminicola* transformed with homologous and heterologous Benomyl-resistance genes retains expected pathogenicity on corn, *Mol. Plant-Microbe Interact.*, 1:113, 1988.

46. Brooker, N. L., Mischke, C. F., Mischke, S., and Lydon, J., Transformation of *Colletotrichum* spp. with the acetylase gene (bar) from *Streptomyces*, in Program of Abstracts of 17th Fungal Genetics Meeting, Asilomar, CA, March 23 to 28, 1993.

47. Brooker, N. L., unpublished data, 1993.

48. Sands, D. C., personal communication, 1993.

49. Bej, A. K., Perlin, M. H., and Atlas, R. M., Model suicide vector for containment of genetically engineered microorganisms, *Appl. Environ. Microbiol.*, 54:2472, 1988.

50. McCormick, D., Detection technology: the key to environmental biotechnology, *Bio/Technology*, 4:419, 1986.

51. Sands, D. C., Miller, R. V., and Ford, E. J., Biotechnological approaches to control of weeds with pathogens, in *Microbes and Microbial Products as Herbicides*, Hoagland, R. E., Ed., American Chemical Society, Washington, D.C., 1990, p. 184.

52. Pannacione, D. G., personal communication, 1993.

53. Miller, R. V., Ford, E. J., and Sands, D. C., A non-sclerotial pathogenic mutant of *Sclerotinia sclerotiorum*, *Can. J. Microbiol.*, 35:517, 1989.

54. Miller, R. V., Ford, E. J., Zidack, N. J., and Sands, D. C., A pyrimidine auxotroph of *Sclerotinia sclerotiorum* for use in biological weed control, *J. Gen. Microbiol.*, 135:2085, 1989.

55. Cove, D. J., Chlorate toxicity in *Aspergillus nidulans*: the selection and characterization of chlorate resistant mutants, *Heredity*, 36:191, 1976.

56. Correll, J. C., Klittich, C. J. R., and Leslie, J. F., Nitrate nonutilizing mutants of *Fusarium oxysporum* and their use in vegetative compatibility tests, *Phytopathology*, 77:1640, 1987.

57. Brooker, N. L., Leslie, J. F., and Dickman, M. B., Nitrate nonutilizing mutants of *Colletotrichum* and their use in studies of vegetative compatibility and genetic relatedness, *Phytopathology*, 6:672, 1991.

58. Andow, D. A., Levin, S. A., and Harwell, M. A., Evaluating environmental risks from biotechnology: contributions of ecology, in *Application of Biotechnology*, Fowle, J.R., III, Ed., American Association for the Advancement of Science, Washington, D.C., 1987, p. 125.

59. Bourquin, A. and Seidler, R., Research plan for test methods development for risk assessment of novel microbes released into terrestrial and aquatic ecosystems, in *Biotechnology and the Environment, Research Needs*, Omenn, G. S. and Teich, A. H., Eds., Noyes Data Corporation, Park Ridge, New Jersey, 1986, p. 18.

60. Fry, W. E., Yoder, O. C., and Apple, A. E., Influence of naturally occurring marker genes on the ability of *Cochliobolus heterostrophus* to induce field epidemics of Southern corn leaf blight, *Phytopathology*, 74:175, 1984.

61. Brill, W. J., Safety concerns and genetic engineering in agriculture, *Science*, 227:381, 1985.

62. Greaves, M. P., Bailey, J. A., and Hargreaves, J. A., Mycoherbicides: opportunities for genetic manipulations, *Pestic. Sci.*, 26:93, 1989.

63. Fuxa, J. R., Environmental risks of genetically engineered entomopathogens, in *Safety of Microbial Insecticides*, Laird, M., Lacey, L. A., and Davidson, E. W., Eds., CRC Press, Boca Raton, Florida, 1990, p. 181.

64. Ohno, S., Immortal genes, *Trends Genet.*, 1:196, 1985.

65. Van der Plank, H. D., *Disease Resistance in Plants*, Academic Press, New York, 1963.

66. Beringer, J. E. and Hirsch, P. R., Genetic adaptation to the environment, the role of plasmids in ecology, in *Current Perspectives in Microbial Ecology*, Klug, M. J. and Reddy, C. A., Eds., American Society for Microbiology, Washington, D.C., 1984, p. 63.

67. Clarke, P. H., Evolution of new phenotypes, in *Current Perspectives in Microbial Ecology*, Klug, M. J. and Reddy, C. A., Eds., American Society for Microbiology, Washington, D.C., 1984, p. 71.

68. Freter, R., Factors affecting conjugal plasmid transfer in natural bacterial communities, in *Current Perspectives in Microbial Ecology*, Klug, M. J. and Reddy, C. A., Eds., American Society of Microbiology, Washington, D.C., 1984, p. 105.

69. Nargang, F. E., Fungal mitochondrial plasmids, *Exp. Mycol.*, 9:285, 1985.

70. Reanney, D. C., Extrachromosomal elements as possible agents of adaptation and development, *Bacteriol. Rev.*, 40:552, 1976.

71. Reanney, D. C., Macphee, D. G., and Pressing, J., Intrinsic noise and the design of the genetic machinery, *Aust. J. Biol. Sci.*, 36:77, 1983.

72. Slater, J. H., Genetic interactions in microbial communities, in *Current Perspectives in Microbial Ecology*, Klug, M. J. and Reddy, C. A., Eds., American Society for Microbiology, Washington, D.C., 1984, p. 87.

73. Carroll, J. M., *Managing Risk, a Computer-Aided Strategy*, Butterworths, Boston, 1984.

74. Nutter, F. W., Assessing the benefits associated with planned introductions of genetically engineered organisms, *Phytopathology*, 81:344, 1991.

75. Levin, S. A. and Harwell, M. A., Potential ecological consequences of genetically engineered organisms, in *Potential Impacts of Environmental Release of Biotechnology Products: Assessment, Regulation, and Research Needs*, Gillett, J. W., Ed., Ecosystems Research Center, Cornell University, Ithaca, New York, 1985, p. 133.

76. Templeton, G. E. and Heiny, D. K., Improvement of fungi to enhance mycoherbicide potential, in *Biotechnology of Fungi for Improving Plant Growth*, Whipps, J. M. and Lumsden, R. D., Eds., Cambridge University Press, London, 1989, p. 127.

77. Turgeon, G. and Yoder, O. C., Genetically engineered fungi for weed control, in *Biotechnology: Applications and Research*, Cheremisinoff, P. N. and Ouellette, R. P., Eds., Technomic Publishing Company, Lancaster, Pennsylvania, 1985, p. 221.

Chapter 8

ENGINEERED PLANTS IN THE ENVIRONMENT

Lidia S. Watrud, Sally G. Metz, and David A. Fischhoff

TABLE OF CONTENTS

1 HISTORICAL PERSPECTIVE ON CROP INTRODUCTIONS, IMPROVEMENTS, AND ESCAPES

Most of the world's major staple food crops — corn, wheat, rice, barley, potatoes, and soybeans — have been intentionally and successfully introduced into geographies, climates, and habitats diverse from their centers of origin. As summarized in 1988 by Kloppenburg,[1] corn originated in Central and South America, wheat and barley in the Mediterranean basin, potatoes and their close relatives, tomatoes in the Andes of South America, rice in Indo-Burma, and soybeans in China. Despite wide-scale geographic introductions and subsequent breeding to increase productivity, disease resistance, insect resistance, and other desired traits, those few agronomic introductions of cultivars of food or feed crops (e.g., oats, sorghum, sugar cane, and rice) which have resulted in localized ecological disturbances can be attributed to weediness of the hybrids which have resulted from outcrossing between the introduced crops and their wild relatives.[2-4] Explanations offered for the limited natural spread of introduced crops per se include requirements for high inputs of fertilizer and water, poor competition with weeds and native vegetation, or absence of weedy relatives in the areas in which the crops are grown.

In contrast to largely annual agricultural food crops, introductions of forage legumes, grasses, and perennial horticultural species have resulted in numerous invasive escapes. As summarized in 1980 by Williams,[5] intentional agronomic introductions which have become weeds include johnsongrass, Bermuda grass, reed canary grass, kudzu, hemp, sicklepod vetch, and tansy. Horticultural ornamental species noted by Williams in the same 1980 review which have become weeds include water hyacinths, Melaleuca, foxglove, and Japanese honeysuckle. Two notable, more recent horticultural introductions which have become noxious weeds, particularly in aquatic environments, are purple loosestrife and Chinese waterspinach.[3]

In addition to physical factors such as light and moisture requirements, soil pH, and fertility, biological factors including symbiotic requirements, disease, insect pressures, and resistances may each be expected to affect the survival, persistence, and potential spread of species that have been intentionally or unintentionally introduced into new environments. Plant characteristics which may be expected to affect their persistence, competitiveness, fate, and ecological effects include (a) whether they are annuals (as are many crop plants) or perennials (as are many of the forage and ornamental species which have become weeds); (b) whether they can reproduce asexually, as well as sexually; (c) length of time pollen or seeds remain viable; (d) the distances and modes by which pollen and seeds may be disseminated; and (e) the activities of specific genes that have been introduced into their genetic complement. Publications which discuss actual consequences or potential risks of introductions are numerous. The selected references highlight examples of consequences[4,6-13] and perceived risks[9,14-24,162] of plant introductions.

2   RATIONALE FOR PLANT GENETIC ENGINEERING

The use of hybrids, modern fertilizers, agricultural chemicals, and other improved cultural practices is widely acknowledged to have boosted yields significantly in the last half-century. However, those gains are counterbalanced in part by the relentless pressure of new strains, races, ecotypes, or genotypes of bacteria, fungi, insects, and weeds that have or may become resistant to chemical bactericides, insecticides, fungicides, or herbicides.[25-29] Concurrently, increasing public and regulatory concerns over the persistence and transport of agricultural chemicals in the environment and their potentially deleterious effects to wildlife, beneficial insects, and humans have resulted in increased interest in crops grown with fewer or more benign agricultural chemical inputs and in biological means of controlling insects, diseases, and weeds.

Plant genetic engineering methods have the potential to provide novel means of accelerating the introduction of traditionally desired agronomic traits such as enhanced pest resistance within or even beyond traditional breeding barriers and in this way capture the benefits of biologically based pest control. They also may provide the means to create novel species and to create unique traits for agronomic pest control, for improved nutrient food processing or storage qualities, for specialty chemical or pharmaceutical production, or for environmental cleanup uses. The impressive variety of old and new applications of plants has attracted a great deal of commercial, governmental, and academic interest. Publications which describe a broad potential variety of agricultural biotechnology applications include the 1990 text on agricultural biotechnology by Lindsey and Jones[30] and the 1992 articles by Fraley[31] and Moffat.[32]

3   MAJOR PLANT GENETIC ENGINEERING METHODS

Two major methods stand out in the evolution of plant genetic engineering: *Agrobacterium tumefaciens Ti* vector-based systems[33-38] and the use of what has been called "particle gun" or "biolistic" technology. The basis and general applications of the *Agrobacterium* and particle gun methods are discussed below in Sections 3.3.1 and 3.3.2, respectively.[39-44]

3.1   *Agrobacterium*-Based Systems

Agrobacterium tumefaciens, a soilborne plant pathogen, is the causal agent of crown gall disease of dicotyledonous plants. In nature, *A. tumefaciens* enters via wounds, generally at or near the soil line or crown. Wounds are caused by implements or insects and by pruning. Once the pathogen has entered its plant host, a segment of DNA called *T*-DNA, derived from an endogenous plasmid called *Ti*, randomly inserts into a chromosome of the host plant, where it provides genetic instructions to produce auxin, a hormone which results in softening of cell walls, plant cell multiplication and enlargement, and subsequent production of a primary tumor or gall. The latter is produced typically near the soil line or "crown" of a plant, hence the name "crown gall" disease.

The attraction of plant biotechnologists to *A. tumefaciens* is its ability to insert DNA into host plant genetic material. Therefore, if a way were developed to "tame" the pathogen, i.e., utilize its integration characteristic, while inactivating its disease-causing capability, the *Ti* plasmid could be used to insert desired genetic information without causing disease, i.e., gall formation typical of its wild-type parent. To that end, the tumor-causing *onc* genes were removed from the *Ti* vector to create, as described by Fraley et al.,[34] a "disarmed" vector system for use in plant transformation. In these *Agrobacterium* systems, the ability to transfer *T*-DNA has been left intact, but the *onc* genes have been replaced by genes of interest to the plant biotechnologist. Specific applications of disarmed *Agrobacterium*-based systems to introduce genes for crop protection, food quality, and specialty uses are discussed below in Section 3.4.

3.2   "Particle Gun" Methodologies

As reviewed in 1992 by Fraley,[31] numerous plant species have already been transformed and interest continues in transforming additional ones for diverse applications. For many dicotyledonous herbaceous and woody species, *Agrobacterium*-based systems have been used for the transformations. With the advent of microprojectile or "particle gun" ballistics-based technologies, transformations of numerous monocotyledonous cereal species such as corn, rice, and wheat have also been reported.[41,45-48] The general basis for particle gun technology is the coating of DNA onto particles of metals such as gold, tungsten, or platinum, which then are accelerated sufficiently to penetrate into host cells, protoplasts, leaves, stems, hypocotyls, or embryos.[39-44] Once inside the cell, a portion of DNA diffuses from the coated particle into a recipient nucleus, where it can integrate into a host plant chromosome. As reported in 1988 articles by Christou et al. and McCabe et al.,[49,50] this approach also has been used for soybeans, a dicotyledonous crop that has not been easily transformed with *Agrobacterium*.

Although the efficiency of achieving stable genetic insertions with particle gun approaches is somewhat less than that with *Agrobacterium*-based systems, it sometimes may be useful in a crop such as corn to facilitate direct transformation of advanced or "elite" breeding lines, as described in 1993 by Koziel et al.[47] The practical result and implication of direct transformation of elite or commercial lines, rather than laboratory lines, is that overall breeding times may be shortened. However, because of the large number of transformants which must be screened to identify ones with sufficient levels of gene expression, it sometimes may be more efficient in some crops such as canola to insert genes into laboratory lines that are more amenable to transformation and to then backcross selected transformed lines to commercial breeding lines.

4 USES OF GENETICALLY ENGINEERED PLANTS IN AGRICULTURE

4.1 Crop Protection

4.1.1 Insect Resistance

The *cry* or delta-endotoxin crystal protein genes from *Bacillus thuringiensis* (*B.t.*) encode highly active, linear polypeptide, insecticidal proteins.[51-53] In the guts of susceptible insects, perhaps after proteolysis, the *cry* proteins bind to receptors on midgut epithelial cells, resulting in disruption of metabolism, probably through the formation of ion-specific channels. The most immediate effects are gut paralysis and cessation of feeding, which can occur in minutes. Death occurs, generally from within a matter of hours to 1 day, either by starvation or by lysis of the insect gut lining.

The *B.t.* genes currently most widely used by biotechnologists to create insect-resistant plants are lepidopteran-active genes isolated from *B.t. kurstaki* or related subspecies. Lepidopteran-resistant plants expressing *B.t. kurstaki* genes have already been field tested in tobacco, tomato, cotton, and corn.[47,54-57] Given the worldwide economic importance of lepidopteran insect pests to food and fiber crops and to silvicultural and horticultural species, interest in transforming many additional herbaceous and woody crops may be anticipated as efficacy is demonstrated and environmental concerns are addressed.

A gene from *B.t. tenebrionis* which encodes an endotoxin which is active against some coleopteran insects such as the Colorado potato beetle was isolated in 1988 by McPherson et al.[58] Early field tests of the efficacy of the gene cloned into potatoes have shown a very high level of beetle control in the field, as noted in a 1992 article by Lundstrum[59] and one in 1993 by Wyman.[60]

Concerns have been raised that insects will develop resistance to biological insecticides expressed in plants, as they have to formerly effective chemical insecticides[25,26,29] and, as described in 1988 by McGaughey and Berman[61] and in 1992 by McGaughey and Johnson,[62] to the microbial pesticide *B.t. kurstaki*. In response to such concerns, entomologists, plant breeders, and biotechnologists are developing and testing methods to incorporate the use of engineered insect-resistant plants into integrated pest management (IPM) systems. A major objective of such IPM-based approaches would be to protect the utility and durability of *B.t.*-expressing plants as pest control agents and to delay or prevent the potential development of insect resistance to *B.t.* proteins expressed in engineered plants. It is expected that development and testing of these IPM strategies will be a major component of future field tests. A 1992 U.S. Department of Agriculture (USDA) workshop on management of resistance to *B.t.*[63] highlighted areas of research needed to address those concerns. These included evaluating the use of refugia or trap plants, using multiple genes for resistance, and using blends of engineered and nonengineered seed in given fields.

The gene for the trypsin inhibitor from cowpeas, described in 1987 by Hilder et al.,[64] is also being evaluated for use in insect control. It has been cloned into tobacco plants and has been tested for its efficacy in insect control. MacIntosh et al.[65] evaluated serine protease inhibitors in combination with the *B.t. kurstaki* delta-endotoxin gene and reported some potentiation of *B.t.* activity by the enzymes. However, serine proteases alone did not appear to bring about a commercially viable level of insect control in engineered plants. Additional sources of novel insecticidal genes are being sought not only from microbes and plants; as reported in 1991 by Tomalski and Miller[66] and in 1991 by Stewart et al.,[67] they are being sought from invertebrates such as mites and scorpions as well.

4.1.2 Disease Resistance

The major strategy being evaluated for engineered resistance to plant viruses is the use of viral coat proteins.[68-70] As reported in 1988 by Hemenway et al.[71] and in 1991 by Moffat,[72] the use of antisense RNAs is also being evaluated. The most common method and the most advanced, which appears to work for a number of RNA viruses in a variety of crop species, has been to clone genes coding for subunits of the coat protein of RNA viruses. Using the coat protein method, high levels of resistance to the mosaic viruses TMV and ToMV in tobacco and tomato, AMV in alfalfa, PVX and PVY in potato, and CMV and ZYMV in squash and melons have been demonstrated in a number of greenhouse and field experiments.[73-79] When challenged with higher titers of inoculum, as reported in 1988 by Hemenway et al.,[71] the antisense approach appeared to confer less protection than the coat protein method for viral disease control. The applicability of ribozyme-based methods designed to cleave viral RNAs, as described in 1987 by Cech,[80] for use in transformed plants is not yet clear.

For fungal and nematode disease control, plant and microbial chitinase genes, ribosome inactivating proteins, and even antibody approaches are being evaluated.[81-85] In the chitinase approach, genes encoding plant or microbial chitinases are cloned into plants to inhibit the growth of chitin-containing fungal or nematode pathogens, as reported in 1992 by Bakker et al.[85] In the antibody approach, genes are being cloned into plants which encode antibodies to chitin associated with fungal or nematode pathogens. Investigation of the efficacy of insect-derived cecropin genes for control of the bacterial diseases such as soft rot of potato was reported in 1987 by Jaynes et al.[86] and in 1991 by Destefano-Beltran et al.[87] An additional approach to bacterial plant disease control, the use of thionins, was reported in 1993 by Carmona et al.[88] Genetic engineering to express plant-derived antimicrobial secondary metabolites, i.e., phytoalexins such as chalcones, or pathogenesis related *PR* proteins, to confer disease resistance, was discussed by Doerner et al.[89] in 1990 and by Moffat[90] in 1992.

4.1.3 Herbicide Tolerance

The agronomic rationale for herbicide-tolerant plants includes improved weed control, decreased chance of crop injury, and increased crop yields. The biochemical bases for engineering herbicide tolerance may differ according to the particular chemical classes of given herbicides.[91-93] They also may differ with regard to the use of microbes or plants as sources of genes being used to confer tolerance. The two approaches most widely utilized for engineering herbicide tolerance are (a) engineering target-site insensitivity, so that the plant target enzyme is rendered less sensitive to the herbicide; and (b) metabolic inactivation, in which the herbicide molecule is converted, perhaps degraded, into an inactive form. Examples of each of these two major approaches are described below.

For tolerance to the nonselective herbicide glyphosate, the target-site method has been tested with both microbial genes and plant genes.[94-98] With both types of gene sources, the genes tested have encoded modified forms of the enzyme 5-enolpyruvyl 3-phosphoshikimate phosphate (EPSP) synthase. These genes encode EPSP enzymes with reduced affinity for binding glyphosate, thereby blocking its herbicidal activity. Major crops for which glyphosate tolerance is being developed and evaluated include soybeans, cotton, and oilseed rape.

For tolerance to sulfonyl ureas, as reported in 1988 by Lee et al.,[99] a target-site approach is being evaluated in which herbicide-tolerant genes for acetolactate synthase (ALS) have been isolated from tobacco.

In another class of herbicides, the bromoxynils, the metabolic inactivation or degradation approach has been utilized. As described in 1988 by

Stalker et al.,[100] a gene cloned from a soil microbe, *Klebsiella*, is being used to confer tolerance to the herbicide in cotton and canola by encoding the enzyme nitrilase, which degrades the herbicide.

In another example of metabolic inactivation or degradation, the *bar* gene for phosphinothricin acetyl transferase was cloned from a soil actinomycete to confer tolerance to the herbicide glufosinate.[91-93]

In addition to tolerances to single herbicides, some genetic engineering or breeding efforts may possibly be anticipated to confer tolerance to combinations of herbicides that might be used for a given crop, e.g., glyphosate with sulfonyl urea for corn, and glyphosate and bromoxynil for cotton.

4.1.4 Multiple Traits for Crop Protection

Efforts to introduce multiple traits of interest into a crop, either by engineering vectors with multiple traits or by sequential engineering and breeding efforts, might be expected as more genes of interest become available. For example, in potatoes, broad-spectrum chemical insecticides currently used to control aphids which transmit potato viruses also may control Colorado potato beetles, wireworms, tuberworms, and nematodes. Similarly, in cotton, currently used chemical insecticides may typically control several types of insects and possibly mites and spiders as well. Thus, since many current insecticide treatments may have numerous pest control benefits, to remain truly competitive in the marketplace, crops initially engineered for one pest control trait may ultimately be expected to be bred or engineered to contain multiple protective traits. To that end, as reported in 1990 by Lawson et al.[76] and by Kaniewski et al.,[77] efforts to transform potato have included development of vectors to confer multiple viral resistances. Ultimately, a combination of Colorado potato beetle, multiple viral resistances, and quality traits such as those discussed below might be anticipated as well.

Depending on the ease of application and the time and costs required, introduction of crop protection and other desired traits into given crop plants may be achieved by combinations of both traditional and genetic engineering breeding methods.

4.2 FOOD QUALITY AND SPECIALTY USES

4.2.1 Control of Ripening and Post-Harvest Physiology

Two major antisense approaches have been used to date to control ripening. In one, antisense genes are being used to suppress the activity of tissue-softening enzymes such as polygalacturonases, as reported in 1988 by Hiatt et al.[101] In the other, as described in 1991 by Oeller et al.,[102] antisense genes to aminocyclopropane synthase block the production of ethylene, the key plant hormone involved in ripening. As reported by Kramer et al. in 1990 and 1992,[103,104] field testing and post-harvest evaluation of tomatoes containing antisense genes to polygalacturonase has already commenced; engineering of additional fruits and vegetables using antisense technology is likely if the approach taken with tomato is successful commercially.

In crops where negative medical connotations may be associated with a secondary metabolite, such as caffeine in coffee, antisense technology potentially could provide a means to produce "safer" products (e.g., naturally decaffeinated coffee). Similarly, plant sources of other foods or beverages potentially could be engineered to have reduced levels of medically problematic constituents. Broad applications of antisense technology are highlighted in a 1991 article by Moffat.[72]

4.2.2 Increased Solids

As reviewed in 1992 by Fraley,[31] two related, usually interconvertible classes of solids — starches and sugars — are the focus of attention in several food crops, including potatoes, tomatoes and corn. Objectives of solids modifications are diverse; they include increasing the content of solids such as starches in potato tubers or corn kernels to reduce oil absorption and excess browning during cooking and increasing the levels of sugars in tomatoes. Specific approaches being taken to modify the amounts and ratio of sugars and specific types of starches in crops such as potatoes and corn include engineering to either block or increase the expression of ADP-glucose pyrophosphorylase enzymes or branching enzymes.[105-107] In 1991, Worrel et al.[108] reported that introduction of a maize phosphate synthase gene altered the partitioning of sugars and starches in tomato leaves.

4.2.3 Enhanced Nutrient Quality

Amino acid supplements today are routinely supplied to the diets of chickens and swine to provide nutrients such as methionine or lysine.[109-111] Such supplements potentially could be replaced or augmented by the use of ground seed or other parts from crops such as soybeans, corn, cotton, alfalfa, or others which have been engineered to contain higher levels of methionine, lysine, or other essential amino acids or vitamins.

A major approach that is being evaluated to improve nutrient quality in soybeans, corn, and other crops is manipulation of the genes for sulfur-rich seed storage proteins isolated from the crops themselves or from other sources, such as the Brazil nut.[112-115]

4.2.4 Increased Tolerance to Environmental Stresses

Tolerance to extremes of heat, cold, or drought and to acid, alkaline, or saline soils could conceivably be useful to a large number of agronomic, silvicultural, and horticultural species. Less commercial activity is apparent so far for alleviation of stress-related problems as compared to engineering for pest control. Technically, the biggest barriers are limitations in knowledge of the genetic or biochemical bases for tolerance to stress phenomena and the likelihood that stress responses may involve multiple genes. In recent years, an increasing number of genes and strategies have been investigated to address plant stress alleviation needs. One gene, an antifreeze protein from fish, was reported in 1990 by Georges et al.[116] and 1991 by Hightower et al.[117] to have been cloned into plants for evaluation for efficacy in tolerance to cold. Bacterial genes for cold tolerance are also being investigated, as reported in 1992 by Baertlein et al.[118] In a 1993 article by Tarczynski et al.,[119] a gene for overproduction of the osmolyte mannitol was shown to confer some salt tolerance to tobacco. Interest in introducing trehalose-synthesizing enzymes into plants to confer drought tolerance was noted in a 1993 article by Danheiser.[120]

4.2.5 Environmental Clean-Up and Restoration

In the area of environmental remediation of soils, sediments, and waters contaminated by pollutants such as heavy metals, solvents, pesticides, and oils, most work is currently centered on the use of civil engineering, microbial bioreactor, land farming, or composting methods, as summarized in the 1992 text by King et al.[121] and in the 1992 U.S. Environmental Protection Agency (USEPA) document on bioremediation of hazardous wastes.[122] That is, contaminated soils might be "washed", incinerated, extracted, or made into slurries that are then treated with microbes in bioreactors. Alternatively, the soils may be "land-farmed", i.e., formed into windrows for further microbial degradation or composting, or treated in situ. Either as a supplement or follow-up to such cleanup practices or, in some instances, where contaminant levels are low enough and the area too large to be amenable to civil engineering or bioreactor treatments, the use of plants is being considered for the cleanup, restoration, and revegetation of contaminated sites and damaged habitats.[123-125] To enhance such applications, efforts have already been initiated to clone into plants metal-binding metallothionein genes from sources as diverse as mice and men, as reported in 1989 by Misra and Gedamu[126] and by Maiti et al.[127] In 1992, Ortiz et al.[128] reported the cloning of yeast-derived membrane-associated phytochelatin genes and noted interest in introducing such genes into plants. Tissue-specific overproduction of phytochelatins in plants was proposed in the same 1992 article by Ortiz et al. as a potential means to sequester heavy metals such as lead and cadmium away from consumable parts of crop plants. For sites having heavy-metal contamination, i.e., areas downwind from metal smelters or surface mining and outdoor ore processing areas, metal-accumulating plants could prove to be particularly useful to minimize soil and water pollution, for cleanup, and for recycling of valuable metals. They may also prove to be of benefit for recovery of radionuclides from contaminated soils.

A variety of metabolic genes used to confer tolerance in crop plants to pesticides such as herbicides, fungicides, or insecticides conceivably also might be useful for environmental cleanup. For example, noncrop plant species could be developed specifically for degrading specific pesticides. Such plants would be useful to remediate and revegetate outdoor areas where chemicals have been stored, distributed, or spilled, with resultant contamination of soil or water. Similarly, noncrop plants engineered for specific degradative or synthetic purposes could potentially be used in the vicinity of chemical manufacturing facilities for pollution prevention and for the recovery and synthesis of desired compounds.

4.2.6   Horticultural and Silvicultural Applications

Flowers of "pure white" or of desired natural or novel colors could theoretically be achieved as an expected result of antisense or sense technologies. To date, floral colors have been achieved both as an expected result of a gene insertion, as reported in 1985 by Meyer et al.,[129] and as initially unexpected effects of gene insertion, as reported in 1990 by MacKenzie[130] and in 1992 by Lloyd et al.[131]

Engineering tolerance to insects, disease, and crop chemicals is technically possible in horticultural species, as it has been in agronomic crops. However, actual development, field testing, and commercialization of engineered horticultural species is limited and appears to be currently focused on high cash crops such as tomatoes and melons. As the costs of developing and accessing the technology go down, one might expect to see an increasing number of horticultural species being engineered with genes of interest. Specific biotechnology applications for floral and woody species are summarized in 1991 articles by Woodson, Scorza, Strauss et al., and Michler.[132-135]

Given the long time — perhaps 25 to 50 years or more — between harvests and the need for replanting of forest species, commercial activity for tree species is currently limited. With the advent and accessibility of *Agrobacterium* and particle gun technologies and the subsequent ease of using traditional breeding to introduce the traits into other lines, more and more tree, nut, and vine species might be expected to contain engineered traits. Use of faster-growing hybrids or varieties of engineered trees which will allow harvests in fewer years conceivably could also increase commercial interest in genetic engineering of tree species.

4.2.7   Specialty Chemical Production

Efforts are slowly being initiated in this as yet largely undeveloped area. Early interests include modification of oils, e.g., of canola and soybeans, to produce higher levels of lauric acid. The latter can then be used as alternatives or replacements for the tropical palm and coconut oils currently used in the manufacture of shampoos, soaps, detergents, and cosmetic products. Technical approaches to modifying oils include the use of "sense" or antisense technology to increase or decrease, respectively, the production of desired intermediate to long-chain oils or fatty acids and to manipulate the desired degree of saturation of carbon-carbon bonds.[136-139]

Perhaps one of the most unique compounds to date being produced in plants via gene transfer, as reported in 1992 by Poirier et al.,[140] is beta-hydroxypolybutyric acid. Although granules of the compound are naturally found in bacteria, expression in plants is significant, since it represents the potential to produce an industrial plastics precursor in a non-petroleum-based system, i.e., crop plants.

As seeds from plants developed for specialty chemical production are obtained, systems may have to be developed to keep specialty seeds separated from those intended for food uses. This separation will be necessary throughout their agronomic culture and subsequent processing. Given the likely higher value of the specialty seeds, one might anticipate the dedicated use of fields or greenhouses as well as processing and storage facilities distinct from those used for food purposes.

4.2.8   Pharmaceutical Production

Recombinant vaccines, hormones, antibiotics, antibodies, and other biologicals currently are produced under rigidly controlled, largely aseptic conditions. Although it is likely that crops can be engineered to produce pharmaceuticals, the economics of extraction and purification of compounds from plants grown in the field will have to be carefully compared with current microbial fermentation, chick embryo, or animal cell tissue culture practices to determine if they are competitive in cost and quality. In addition, any potential environmental and medical concerns and liabilities of exposures of people, wildlife, and nontarget species to pollen and other plant parts will also have to be carefully considered.

To date, numerous experimental demonstrations have been made for production of prototype pharmaceuticals in engineered plants. These include enkephalins, immunoglobulins, alpha-interferon, and alkaloids.[141-144] Recently, attempts have also been made to produce animal and virus antibodies in plants.[145-147]

5 ENVIRONMENTAL ISSUES

The proposed regulatory framework in which to address environmental issues that have been raised for genetically engineered plants, particularly pesticidal ones, is similar to existing regulations for the release of chemical pesticides and microbial pesticides under the Federal Insecticide, Fungicide, and Rodenticide Act (FIFRA): that is, how does one identify the product, determine its environmental fate, and assess its ecological effects? Each of these broad categories — identity, fate, and effects — is briefly summarized below in the context in which it has been discussed at various symposia held between 1985 and 1992 which dealt with risk assessment of genetically engineered plants.[2,148-150,162] Consensus conclusions one may draw from those meetings are the following: (a) risk is driven largely by the source and stability of the introduced gene, the activity of gene product, and the biology of its intended host plant; and (b) risk assessments for nontarget ecological and health effects and food safety should generally be considered on a case-by-case basis. Based largely on such thinking, the USDA, the USEPA, and the United States Food and Drug Administration (USFDA) have all proposed regulatory frameworks in which to deal with engineered plant products. Several plant products are now moving through these systems.

5.1 Identification and Characterization of Genetically Engineered Plant Products

Modern molecular methods such as those based on use of immunological or nucleotide probes can facilitate the isolation, characterization, and tracking of genes introduced into plants and the products they yield in their engineered plant hosts. In addition to characterization of engineered genes, their vectors, and gene products, characterization of the plant host is considered useful in predicting (a) the stability, fate, and ecological effects of the introduced trait, and (b) the fate and effects of the engineered plant itself. Plant characteristics generally considered to play potential roles in the fate, dissemination, persistence, and ecological effects of engineered plants include whether the plant is an annual or perennial, its means of propagation (sexual and/or asexual), whether the pollen is wind or insect pollinated, if wild relatives with which it may hybridize exist in the area, and the length of time the pollen and seeds are likely to remain viable.

5.2 Fate of Engineered Products

As plant genetic engineering has advanced, numerous questions have been raised regarding the environmental fate of genes introduced into plants. For example, will genes from engineered crops transfer to nonengineered related or unrelated crop or weedy species? Will gene exchange, i.e., cross-pollinations, result in viable progeny? Will those progeny persist, and will they be more competitive, invasive, or persistent than their parent crop?

Early data from canola, potato, and cotton field tests suggest that limited outcrossing may be expected, particularly in open-pollinated species.[151-157] However, this may not be the case for crops such as sorghum, oats, sunflowers, clover, and alfalfa, where more extensive outcrossing may be anticipated.[2,158] As the results of more field tests become available and are published, the outcrossing and weediness potential of engineered and nonengineered crops can be better evaluated and compared.

To a much lesser degree than outcrossing and weediness concerns, questions have been raised with regard to the ability of DNA in engineered plants to transform other life forms. That is, is it likely or only a very remote probability that the DNA in pollen or plant parts consumed by insects, earthworms, or vertebrates can transform their gut flora? If the DNA becomes bound to soil particles, will it be biologically active or will it become rapidly degraded? In 1992, Hoffman et al. reported preliminary evidence for horizontal gene transfer between higher plants and the fungus *Aspergillus niger* and cited several additional examples of "transkingdom" gene exchange, including the well-known *Agrobacterium-dicotyledenous* plant interaction.[159]

With regard to the fate of engineered gene products in plants, little work has been published to date. Results reported in 1994 by Palm et al.[160] suggest that the *B.t. kurstaki* delta-endotoxin engineered in cotton leaf tissue degrades relatively rapidly, over a period of several weeks in some soils; this apparent degradation appears to be faster than that for the purified microbial delta-endotoxin protein isolated from *B.t. kurstaki*.

For virus-resistant plants, numerous questions were raised in 1991 by de Zoeten[161] regarding the possibility, frequencies, and effects of interactions with other plant viruses. Viral transencapsidation and recombination phenomena and potential effects were also discussed at the symposium held at the University of Maryland in 1992.[162] Whether or not novel viruses resulting from those phenomena could have expanded, host ranges may have to be determined on a case-by-case basis. Similarly, the viability and fecundity of the insects which vector plant viruses may also be impacted by virus-resistant plants and thus may also need to be considered. The propensity of given types of viruses to recombine more frequently or be more unstable than other types of viruses may also be a risk factor to consider in predicting the ecological consequences of large-scale releases of virus-resistant plants.

5.3   Ecological Effects

Perhaps the most difficult questions to address that are raised with regard to risk assessment of genetically engineered plants relate to the lack of consensus or availability of methods for predicting and testing the short- and long-term effects on nontarget species or communities, ecosystem functioning, and biodiversity.

An ecological issue which frequently is brought up for insect-tolerant crops is the potential for development of resistance by insects to plant-expressed endotoxins.[29,61] This concern has been argued by some to be an issue of crop and product life management rather than a risk assessment issue. To date, development of resistance to chemical pesticides has generally been considered to be an economic and efficacy issue, not a regulatory one. Strategies being developed and tested by academic, government, and industrial scientists to delay or prevent development of resistance include using multiple genes to confer tolerance to given insects, using crop rotation, use of trap crops as refugia, and the use of blends of engineered and nonengineered seed in given fields.[63]

6   FIELD TESTING EXPERIENCE

Hundreds of field tests of genetically engineered plants are estimated to have taken place in the U.S. between 1987 and 1992.[163,164] The exact number of tests or test sites is in question, since estimates have been made using different criteria for what constitutes a test. The number of U.S. field tests has been variously estimated by the number of USDA permits issued and/or by the numbers of states, sites, crops, genes, or constructs for which the permit was requested. Regardless of the criteria used to estimate the number of U.S. field tests (estimates range from 400–600), two messages are clear: commercial and research interest is high and test and permit numbers are increasing at a rapid rate. The total number of tests in Canada, Europe, New Zealand, Australia, and Latin America is also estimated to total in the hundreds. Worldwide, approximately two to three dozen plant species are estimated to have been tested in more than two dozen countries.[165-167]

Table 1 summarizes the types of permits which were requested in the U.S. between 1987 and October 1992 for field testing genetically engineered plants. The largest trait category, representing perhaps one third or more of U.S. tests to date, based on the number of USDA Animal and Plant Health Inspection Service (APHIS) approvals for field testing, was for herbicide tolerance. The next two largest categories were insect resistance and virus resistance, respectively. Plant species tested in the U.S. included tomatoes, cotton, potatoes, soybeans, tobacco, corn, and alfalfa. Major

TABLE 1

U.S. Field Tests 1987–1992

A. Traits	B. Crops
Herbicide tolerane	Soybean
Insect resistance	Potato
Virus resistance	Cotton
Bacterial resistance	Tomato
Fruit development/ripening	Corn
Altered storage protein	Tobacco
Increased solids	Cantaloupe
Fungal resistance	Squash
Seed oil modifications	Alfalfa
Bruising resistance	Rapeseed
Male sterility	Rice
Altered sweetening	Cucumber
Metal chelation	Chrysanthemum
	Melon
	Walnut
	Papaya
	Plum
	Poplar
	Serviceberry
	Apple

Note: Information above is presented in descending order, based on frequency of USDA-APHIS permit requests. Summary information for table was supplied courtesy of Jane Rissler, National Wildlife Federation.

crops tested in Canada included canola and potatoes. Oilseed rape, sugarbeets, and potatoes were among the major crops tested in Europe.[165-167]

In 1993, USDA-APHIS proposed, via a notice in the *Federal Register*, that field testing of certain traits in selected crops in the U.S. (potatoes, tomatoes, corn, soybeans, cotton, and tobacco) be put on a notification rather than on a permitting basis.[168] For given types of crops and specified types of traits, a 30-day notification to APHIS prior to field testing could suffice; i.e., permits would not be required prior to small-scale field testing. However, applications for novel genes and novel applications would continue to be considered on a case-by-case basis.

7   PROSPECTS

As this text comes to press, commercial interest remains high as field and laboratory demonstrations of efficacy of the technology continue to be largely encouraging. Issues of gene and germplasm ownership, profits, and cost of access to the fruits of technology continue to be topics of discussion.[169-172] Questions regarding the economic and biological effects of large monocultures of engineered crops on crop and pest evolution and biological diversity also will likely be discussed and studied. Efficacy, availability, cost, ecological effects, and public opinions will influence which products will be accepted by seed companies, growers, food processors, and end consumers. As the costs of development go down and environmental questions are increasingly addressed, worldwide use of engineered plants might be anticipated. To help address long-term environmental concerns, continued publication and analysis of field monitoring and risk assessment data will be useful to help ensure the environmental as well as product safety and efficacy of engineered plant products.

REFERENCES

1. Kloppenburg, J. R., *First the Seed — the Political Economy of Plant Biotechnology 1492–2000*, Cambridge University Press, London, 1988, 349.
2. Boyce Thompson Institute for Plant Research, Regulatory Considerations: *Genetically Engineered Plants*, Summary of a workshop held at Cornell University, October 19–21, 1987, Center for Science Information, San Francisco, CA, 1988.
3. Westbrooks, R., Personal communication, USDA-APHIS, Plant Protection and Quarantine Unit, Whiteville, NC, 1993.
4. DeWet, J. M. J. and J. R. Harlan, Weeds and domesticated evolution in the man-made habitat, *Econ. Bot.*, 29, 99, 1975.
5. Williams, M. C., Purposefully introduced plants that have become noxious or poisonous weeds, *Weed Sci.*, 28, 300, 1980.
6. Baker, H. G., The evolution of weeds, *Annu. Rev. Ecol. Syst.*, 5, 1, 1974.
7. Dale, P. J., Spread of engineered genes to wild relatives, *Plant Physiol.*, 100, 13, 1992.
8. Foy, C. L. and D. R. Forney, A history of the introduction of weeds, in *The Movement and Dispersal of Agriculturally Important Biotic Agents*, MacKenzie, D. R., Barfield, C. S., Kennedy, G. G., Berger, R. D., and Taranto, D. J. Eds., Claitor's Publishing Division, Baton Rouge, LA, 1985, 115.
9. Morgenthaler, E., What's Florida to do with an explosion of Melaleuca trees?, *Wall Street Journal*, February 8, 1993.
10. Turner, C., Ecology of invasions by weeds, in *Weed Management in Agroecosystems: Ecological Approaches*, Alteri, M. A. and Liebman, M. Eds., CRC Press, Boca Raton, FL, 1988, 354.
11. Holdgate, M. W., Summary and conclusions: characteristics and consequences of biological invasions, *Philos. Trans. R. Soc. of London*, B314, 733, 1986.
12. Mooney, H., S. Hamburg, and J. Drake, The invasions of plants and animals into California, in *Ecology of Biological Invasions of North America and Hawaii*, Mooney, J. A. and Drake, J. A. Eds., Springer-Verlag, New York, 1986, 250.
13. Williamson, M. H. and K. C. Brown, The analysis and modelling of British invasions, *Philos. Trans. R. Soc. London*, B314, 505, 1986.
14. Kareiva, P., R. Manasse, and W. Morris, Using models to integrate data from field trials and estimate risks of gene escape and gene spread, in *Biological Monitoring of Genetically Engineered Plants and Microbes*, MacKenzie, D. R. and Henry, S. C. Eds., ARI Press, Washington, D.C., 1991, 31.

15. Perrins, J., M. Williamson, and A. Fitter, A survey of differing views of weed classification: implications for regulation of introductions, *Biol. Conserv.*, 60, 47, 1992.

16. Regal, P., Gene flow and adaptability in transgenic agricultural organisms: long-term risks and overview, in *Proceedings of the International Conference on Risk Assessment in Agricultural Biotechnology*, Marois, J. J. and Bruening, G. Eds., University of California, Davis, CA, 1990.

17. Keeler, K., Can genetically engineered crops become weeds?, *Bio/Technology*, 7, 1134, 1989.

18. Beringer, J. E. and M. J. Dale, The release of genetically engineered plants and microorganisms, *J. Chem. Tech. Biotechnol.*, 43, 273, 1988.

19. Ellstrand, N., Pollen as a vehicle for the escape of engineered genes?, *Trends Ecol. Evol. Trends Biol.* (combined issue), 3 and 6, 30, 1988.

20. Ellstrand, N. and C. Hoffman, Hybridization as an avenue of escape for engineered genes, *BioScience*, 40, 438, 1990.

21. Evenhuis, A. and J. C. Zadoks, Possible hazards to wild plants of growing transgenic plants. A contribution to risk analysis, *Euphytica*, 55, 81, 1991.

22. Tiedje, J. M., R. K. Colwell, Y. L. Grossman, R. E. Hodson, R. E. Lenski, R. N. Mack, and P. J. Regal, The planned introduction of genetically engineered organisms: ecological considerations and recommendations, *Ecology*, 70, 298, 1989.

23. Williamson, M., J. Perrins, and A. Fitter, Releasing genetically engineered plants: present proposals and possible hazards, *Trends Ecol. Evol.*, 5, 417, 1990.

24. Williamson, M., Environmental risks from the release of genetically modified organisms (GMOs) — the need for molecular ecology, *Mol. Ecol.*, 1, 3, 1992.

25. Georghiou, G. P. and C. Taylor, Factors influencing the evolution of resistance, in *Pesticide Resistance: Strategies and Tactics for Management*, National Academy Press, Washington, D.C., 1986, 157.

26. Georghiou, G. P., The magnitude of the resistance problem, in *Pesticide Resistance: Strategies and Tactics for Management*, National Academy Press, Washington, D.C., 1986, 14.

27. Georghiou, G. P. and A. Lagunes, The occurrence of resistance to pesticides in arthropods, *Food and Agricultural Organization of the United Nations*, Rome, 1991, 318.

28. Gould, F., A. Martinez-Ramirez, A. Anderson, J. Ferre, F. Silva, and W. Moar, Broad-spectrum resistance to *Bacillus thuringiensis* toxins in *Heliothis virescens*, *Proc. Natl. Acad. Sci. U.S.A.*, 89, 17, 7986, 1992.

29. Gould, F., The evolutionary potential of crop pests, *Am. Sci.*, 79, 496, 1991.

30. Lindsey, K. and M. G. K. Jones, *Plant Biotechnology in Agriculture*, Prentice Hall, Englewood Cliffs, NJ, 1990, 241.

31. Fraley, R., Sustaining the food supply, *Bio/Technology*, 10, 40, 1992.

32. Moffat, A. S., High-tech plants promise a bumper crop of new products, *Science*, 256, 770, 1992.

33. Nester, E. W., M. P. Gordon, R. M. Amasino, and M. F. Yanofsky, Crown gall: a molecular and physiogical analysis, *Annu. Rev. Plant Physiol.*, 35, 387, 1984.

34. Fraley, R. T., S. G. Rogers, B. R. Horsch, P. Sanders, J. Flick, S. Adams, M. Bittner, L. Brand, C. Fink, J. Fry, G. Gallupi, S. Goldberg, N. Hoffman, and S. Woo, Expression of bacterial genes in plant cells, *Proc. Natl. Acad. Sci. U.S.A.*, 80, 4803, 1983.

35. Fraley, R. T., S. G. Rogers, and R. B. Horsch, Genetic transformation in higher plants, *Crit. Rev. Plant Sci.*, 4, 1, 1986.

36. Zambryski, P., Basic processes underlying *Agrobacterium*-mediated DNA transfer to plant cells, *Annu. Rev. Genet.*, 22, 1, 1988.

37. Corbin, D. R. and H. J. Klee, *Agrobacterium tumefaciens* mediated plant transformation systems, *Curr. Opin. Biotechnol.*, 2, 147, 1989.

38. Gasser, C. S. and R. T. Fraley, Genetically engineering crops for crop improvement, *Science*, 244, 1293, 1989.

39. Sanford, J., T. Klein, E. Wolf, and N. Allen, Delivery of substances into cells and tissues using a particle bombardment process, *Partic. Sci. Technol.*, 5, 27, 1987.

40. Klein, T. M., E. D. Wolf, R. Wu, and J. C. Sanford, High velocity microprojectiles for delivering nucleic acids into living cells, *Nature*, 327, 70, 1987.

41. Potrykus, I., Gene transfer to cereals: an assessment, *Bio/Technology*, 8, 535, 1990.

42. Klein, T. M., M. Fromm, A. Weissinger, D. Tomes, S. Schaff, M. Sletten, and J. C. Sanford, Transfer of foreign genes into maize cells with high-velocity microprojectiles, *Proc. Natl. Acad. Sci. U.S.A.*, 85, 4305, 1988.

43. Sanford, J. C., Biolistic plant transformation, *Physiol. Plant.*, 79, 206, 1990.

44. Klein, T. M., R. Arentzen, P. A. Lewis, and S. Fitzpatrick-McElligott, Transformation of microbes, plants and animals by particle bombardment, *Bio/Technology*, 10, 286, 1992.

45. Toriyama, K., Y. Arimoto, H. Uchimiya, and K. Hinato, Transgenic rice plants after direct gene transfer into protoplasts, *Bio/Technology*, 6, 1072, 1988.

46. Christou, P., T. L. Ford, and M. Kofron, Production of transgenic rice (*Oryza sativa* L.) plants from agronomically important indica and japonica varieties via electronic discharge particle acceleration of exogenous DNA into immature zygotic embryos, *Bio/Technology*, 9, 957, 1991.

47. Koziel, M. G., G. L. Beland, C. Bowman, N. B. Carozzi, R. Crenshaw, L. Crossland, J. Dawson, N. Desai, M. Hill, S. Kadwell, K. Launis, K. Lewis, D. Maddox, K. McPherson, M. R. Meghji, E. Merlin, R. Rhodes, G. R. Warren, M. Wright, and S. V. Evola, Field performance of elite transgenic maize plants expressing an insecticidal protein derived from *Bacillus thuringiensis*, *Bio/Technology*, 11, 194, 1993.

48. Vasil, V., A. M. Castillo, M. E. Fromm, and I. K. Vasil, Herbicide resistant fertile transgenic wheat plants obtained by microprojectile bombardment of regenerable embryogenic callus, *Bio/Technology*, 10, 667, 1992.

49. Christou, P., D. E. McCabe, and W. F. Swain, Stable transformation of soybean callus by DNA-coated gold particles, *Plant Physiol.*, 87, 671, 1988.

50. McCabe, D. E., W. F. Swain, B. J. Martinell, and P. Christou, Stable transformation of soybean (*Glycine max*) by particle acceleration, *Bio/Technology*, 6, 923, 1988.

51. Hofte, H. and H. R. Whiteley, Insecticidal crystal proteins of *Bacillus thuringiensis*, *Microbiol. Rev.*, 53, 242, 1989.

52. Meeusen, R. K. and G. Warren, Insect control with genetically engineeered crops, *Annu. Rev. Entomol.*, 34, 373, 1989.

53. Adang, J. M., *Bacillus thuringiensis* insecticidal crystal proteins: gene structure, action and utilization, in *Biotechnology for Biological Control of Pests and Vectors*, Maramorosch, K. Ed., CRC Press, Boca Raton, FL, 1991, 3.

54. Vaeck, M., A. Reynaerts, H. Hafte, S. Jansens, M. DeBeuckeleer, C. Dean, M. Zakeau, M. VanMontagu, and J. Lermans, Transgenic plants protected from insect attack, *Nature*, 328, 33, 1987.

55. Delannay, X., B. J. LaVallee, R. K. Proksch, R. L. Fuchs, S. R. Sims, J. J. Augustine, J. G. Layton, and D. A. Fischhoff, Field performance of transgenic tomato plants expressing the *Bacillus thuringiensis* var. *kurstaki* insect control protein, *Bio/Technology*, 7, 1265, 1989.

56. Fischhoff, D. A., K. S. Bowdish, F. J. Perlak, P. G. Marrone, S. M. McCormick, J. G. Niedermeyer, D. A. Dean, K. Kusano-Kretzmer, E. J. Mayer, D. E. Rochester, S. G. Rogers, and R. T. Fraley, Insect tolerant transgenic tomato plants, *Bio/Technology*, 5, 807, 1988.

57. Umbeck, P. F., K. A. Barton, E. V. Nordheim, J. C. McCarty, W. L. Parrott, and J. N. Jenkins, Degree of pollen dispersal by insects from a field test of genetically engineered cotton, *J. Econ. Entomol.*, 84, 1943, 1991.

58. McPherson, S. A., F. J. Perlak, R. L. Fuchs, P. G. Marrone, P. B. Lavrik, and D. A. Fischhoff, Characterization of the coleopteran-specific protein gene of *Bacillus thuringiensis* var. *tenebrionis*, *Bio/Technology*, 6, 61, 1988.

59. Lundstrum, L., Monsanto develops beetle resistant plants — plots show remarkable control, *Potato Grower of Idaho*, 21, 3, 36, 1992.

60. Wyman, J., Revolution! Beetle genetics successful in Wisconsin, *Spudman*, 30, 8, 1992.

61. McGaughey, W. H. and R. W. Berman, Resistance to *Bacillus thuringiensis* in colonies of Indian meal moth and almond moth (Lepidoptera:pyralidae), *J. Econ. Entomol.*, 81, 28, 1988.

62. McGaughey, W. H. and D. E. Johnson, Indian meal moth (Lepidoptera: Pyralidae) resistance to different strains and mixtures of *Bacillus thuringiensis*, *J. Econ. Entomol.*, 85, 5, 1594, 1992.

63. USDA-CRS and USDA-ARS, Scientific Evaluation of the Potential for Pest Resistance to the *Bacillus thuringiensis* (*Bt*) Delta-Endotoxins. A conference to explore resistance management strategies, Beltsville, MD, January 21–23, 1992.

64. Hilder, V. A., A. M. R. Gatehouse, S. E. Sheerman, R. F. Barker, and D. Boulter, A novel mechanism of insect resistance engineered into tobacco, *Nature*, 330, 160, 1987.

65. MacIntosh, S. C., G. M. Kishore, F. J. Perlak, P. G. Marrone, T. B. Stone, S. R. Sims, and R. L. Fuchs, Potentiation of *Bacillus thuringiensis* insecticidal activity by serine protease inhibitors, *J. Agric. Food Chem.*, 38, 1145, 1990.

66. Tomalski, M. and L. Miller, Insect paralysis by baculovirus mediated expression of a mite neurotoxin gene, *Nature*, 352, 82, 1991.

67. Stewart, L. M. D., M. Hirst, M. L. Ferber, A. T. Merryweather, P. J. Cayley, and R. D. Possee, Construction of an improved baculovirus insecticide containing an insect-specific toxin gene, *Nature*, 352, 85, 1991.

68. Powell-Abel, P., R. S. Nelson, B. Re, N. Hoffmann, S. G. Rogers, R. T. Fraley, and R. N. Beachy, Delay of disease development in transgenic plants that express the tobacco mosaic virus coat protein gene, *Science*, 232, 738, 1986.

69. Beachy, R. N., S. Loesch-Fries, and N. E. Tumer, Coat protein-mediated resistance against virus infection, *Annu. Rev. Phytopathol.*, 28, 451, 1990.

70. Gadani, F., L. M. Mansky, R. Medici, W. A. Miller, and J. H. Hill, Genetic engineering of plants for virus resistance, *Arch. Virol.*, 115, 1, 1990.

71. Hemenway, C., R. X. Fang, J. J. Kaniewski, N. H. Chua, and N. E. Tumer, Analysis of the mechanism of protection in transgenic plants expressing the potato virus X coat protein or its antisense RNA, *EMBO J.*, 7, 1273, 1988.

72. Moffat, A. S., Making sense of antisense: use of antisense RNA to block gene expression useful, e.g., for disease therapy and plant breeding, *Science*, 253, 510, 1991.

73. Tumer, N. E., K. M. O'Connell, R. S. Nelson, P. R. Sanders, R. N. Beachy, R. T. Fraley, and D. M. Shah, Expression of alfalfa mosaic virus coat protein confers cross-protection in transgenic tobacco and tomato plants, *EMBO J.*, 6, 1181, 1987.

74. Cuozzo, M., K. M. O'Connell, W. Kaniewski, R. X. Fang, N. H. Chua, and N. E. Tumer, Viral protection in transgenic tobacco plants expressing the cucumber mosiac virus coat protein or its antisense RNA, *Bio/Technology*, 6, 549, 1988.

75. Anderson, E., D. M. Stark, R. S. Nelson, P. A. Powell, N. E. Tumer, and R. N. Beachy, Transgenic plants that express the coat protein genes of tobacco mosaic virus or alfalfa mosaic virus interfere with disease development of some nonrelated viruses, *Phytopathology*, 79, 1284, 1989.

76. Lawson, C., W. Kaniewski, L. Haley, R. Rozman, C. Newell, P. Sanders, and N. E. Tumer, Engineering resistance to mixed virus infection in a commercial potato cultivar: resistance to potato virus X and potato virus Y in transgenic Russet Burbank, *Bio/Technology*, 8, 127, 1990.

77. Kaniewski, W., C. Lawson, B. Sammons, L. Haley, J. Hart, X. Delannay, and N. E. Tumer, Field resistance of transgenic Russet Burbank potato to effects of infection by potato virus X and potato virus Y, *Bio/Technology*, 8, 750, 1990.

78. Gonsalves, D., P. Chee, R. Provvidenti, R. Seem, and J. Slighton, Comparison of coat protein-mediated and genetically derived resistance in cucumbers to infection by cucumber mosaic virus under field conditions with natural challenge inoculations by vectors, *Bio/Technology*, 10, 1562, 1992.

79. Nelson, R. S., S. M. McCormick, X. Delannay, P. Dube, J. Layton, E. J. Anderson, M. Kaniewska, R. K. Proksch, R. B. Horsch, S. G. Rogers, R. T. Fraley, and R. N. Beachy, Virus tolerance, plant growth, and field performance of transgenic tomato plants expressing coat protein from tobacco mosaic virus, *Bio/Technology*, 6, 403, 1988.

80. Cech, T. R., The chemistry of self-splicing RNA and RNA enzymes, Science, 236, 1532, 1987.

81. Schlombaum, A., F. Mauch, U. Vogeli, and T. Boller, Plant chitinases are potent inhibitors of fungal growth, *Nature*, 324, 365, 1986.

82. Lund, P., R. Y. Lee, and P. Dunsmuir, Bacterial chitinase is modified and secreted in transgenic tobacco, *Plant Physiol.*, 91, 130, 1989.

83. Broglie, K., I. Chet, M. Holliday, R. Cressman, P. Biddle, S. Knowlton, C. J. Mauvais, and R. Broglie, Transgenic plants with enhanced resistance to the fungal pathogen *Rhizoctonia solani*, Science, 254, 5035, 1194, 1991.

84. Lageman, J., G. Jack, H. Tommerup, J. Mundy, and J. Schell, Expression of a barley ribosome-inactivating protein leads to increased fungal protection in transgenic tobacco plants, *Bio/Technology*, 10, 305, 1992.

85. Bakker, J., A. Schots, W. J. Steikma, and F. J. Gommers, Plantibodies: a versatile approach to engineer resistance against pathogens, in *Proceedings of the 2nd International Symposium on the Biosafety Results of Field Tests of Genetically Modified Plants and Microorganisms, May 11–14*, Goslar, Germany, Casper, R. and Landsmann, J. Eds., Biologische Bundesanstalt für Land- und Forstwirtschaft, Braunschweig, Germany, 1992, 61.

86. Jaynes, J. M., K. G. Xanthopoulos, L. Destefano-Beltran, and J. H. Dodds, Increasing bacterial disease resistance in plants utilizing antibacterial genes from insects, *BioEssays*, 6, 263, 1987.

87. Destefano-Beltran, L., P. Nagpala, K. Jaeho, J. H. Dodds, and J. M. Jaynes, Genetic transformation of potato to enhance nutritional value and confer disease resistance, in *Molecular Approaches to Crop Improvement*, Dennis, E. S. and Llewellyn, D. J. Eds., Springer-Verlag, New York, 1991, 17.

88. Carmona, M. J., A. Molina, J. A. Fernandez, J. J. Lopez-Fando, and F. Garcia-Olmedo, Expression of the α-thionin gene from barley in tobacco confers enhanced resistance to bacterial pathogens, *Plant J.*, 3, 457, 1993.

89. Doerner, P. W., B. Sterner, J. Schmid, R. A. Dixon, and C. J. Lamb, Plant defense gene promoter-reporter gene fusions in transgenic plants: tools for identification of novel inducers, *Bio/Technology*, 8, 845, 1990.

90. Moffat, A. S., Improving plant disease resistance, *Science*, 257, 482, 1992.

91. Mazur, B. J. and S. C. Falco, The development of herbicide resistant crops, *Annu. Rev. Plant Physiol. Plant Mol. Biol.*, 40, 441, 1989.

92. Padgette, S. R., G. della-Cioppa, D. M. Shah, R. T. Fraley, and G. M. Kishore, Selective herbicide tolerance through protein engineering, in *Cell Culture and Somatic Cell Genetics of Plants*, Academic Press, New York, 1989, 441.

93. Hinchee, M. A. W., S. R. Padgette, G. M. Kishore, X. Delannay, and R. T. Fraley, Herbicide-tolerant crops, in *Transgenic Plants: Engineering and Utilization*, Vol. 1, Academic Press, New York, 1993, 243.

94. Comai, L., L. Sen, and G. Stalker, An altered aro A gene product confers resistance to the herbicide glyphosate, *Science*, 221, 370, 1983.

95. Comai, L., D. Facciotti, W. R. Hiatt, G. Thompson, R. Rose, and D. Stalker, Expression in plants of a mutant aro A gene from *Salmonella typhimurium* confers tolerance to glyphosate, *Nature*, 317, 741, 1985.

96. Kishore, G. M. and D. Shah, Glyphosate-tolerant 5-enolpyrunyl 3-phosphoshikimate synthase, *Biotechnol. Adv.*, 9, 1, 1991.

97. Shah, D., R. Horsch, H. Klee, G. Kishore, J. Winter, N. Tomer, C. Hironaka, P. Sanders, C. Gasser, S. Aykent, N. Siegel, S. Rogers, and R. Fraley, Engineering herbicide tolerance in transgenic plants, *Science*, 233, 478, 1986.

98. Barry, G., G. Kishore, S. Padgette, M. Taylor, K. Kolaoz, M. Weldon, D. Re, D. Eichholtz, K. Fincher, and L. Hallas, Inhibitors of amino acid biosynthesis: strategies for imparting glyphosate tolerance to crop plants, in *Biosynthesis and Molecular Regulation of Amino Acids in Plants*, Singh, B. K., Flores, H. E., and Shannon, J. C. Eds., American Society of Plant Physiologists, Rockville, MD, 1992, 139.

99. Lee, K. Y., J. Townsend, J. Tepperman, M. Black, C. F. Chui, B. Mazun, P. Dunsmuir, and J. Bedbrook, The basis of sulfonyl urea herbicide resistance in tobacco, *EMBO J.*, 7, 1241, 1988.

100. Stalker, D. M., K. E. McBride, and D. Malyj, Herbicide resistance in transgenic plants expressing a bacterial detoxification gene, *Science*, 242, 419, 1988.

101. Hiatt, W., M. Kramer, and R. Sheehy, The application of antisense RNA technology to plants, in *Genetic Engineering, Principles and Methods*, Vol. 11, Setlow, J. Ed., Plenum Press, New York, 1989, 49.

102. Oeller, P. W., L. Min-Wong, L. P. Taylor, D. A. Pike, and A. Theologis, Reversible inhibition of tomato fruit senescense by antisense RNA, *Science*, 254, 437, 1991.

103. Kramer, M., R. Sanders, R. Sheehy, M. Melis, M. Kuehn, and W. Hiatt, Field evaluation of tomatoes with reduced polygalacturonase by antisense RNA, in *Horticultural Biotechnology*, Bennett, A. and O'Neil, S. Eds., Wiley-Liss, New York, 1990, 347.

104. Kramer, M., R. Sanders, H. Bolkan, C. Waters, R. Sheehy, and W. Hiatt, Post-harvest evaluation of transgenic tomatoes with reduced levels of polygalacturonase: processing, firmness and disease resistance, *Post Harvest Biol. Technol.*, 1, 241, 1992.

105. Muller-Rober, B., U. Sonnewald, and L. Willmitzer, Inhibition of the ADP-glucose pyrophosphorylase in transgenic potatoes leads to sugar-storing tubers and influences tuber formation and expression of tuber storage protein genes, *EMBO J.*, 11, 1229, 1992.

106. Shewmaker, C. K. and D. M. Stalker, Modifying starch biosynthesis with transgenes in potatoes, *Plant Physiol.*, 100, 1083, 1992.

107. Kossman, J., R. G. Visser, and B. Muller-Rober, Cloning and expression analysis of a potato cDNA that encodes branching enzyme: evidence for co-expression of starch biosynthetic genes, *Mol. Gen. Genet.*, 230, 39, 1991.

108. Worrell, A. C., J.-M. Bruneau, K. Summerfelt, M. Boersig, and T. A. Voelker, Expression of a maize sucrose phosphate synthase in tomato alters leaf carbohydrate partitioning, *Plant Cell*, 3, 1121, 1991.

109. Leeson, L. and J. D. Summers, *Commercial Poultry Nutrition*, University Books, Guelph, Ontario, Canada, 1991, 283.

110. Maynard, L. A., J. K. Looski, H. F. Hintz, and R. G. Warner, *Animal Nutrition*, 7th Ed., Campbell, J. R. and Hall, C. Eds., McGraw-Hill Book Company, New York, 1979, 602.

111. Church, D. C., *Livestock Feeds and Feeding*, 3rd edition, Prentice-Hall, Englewood Cliffs, NJ, 1991, 546.

112. Slighton, J. L. and P. O. Chee, Advances in the molecular biology of plant seed storage proteins, *Biotechnol. Adv.*, 5, 29, 1987.

113. Hoffman, L. M., D. D. Donaldson, and E. M. Herman, A modified storage protein is synthesized, processed and degraded in the seeds of transgenic plants, *Plant Mol. Biol.*, 11, 717, 1988.

114. Kritz, A. L. and B. A. Larkins, Biotechnology of seed crops: genetic engineering of seed storage proteins, *HortScience*, 26, 1036, 1991.

115. Krebbers, E. and J. Vandekerckhove, Production of peptides in plant seeds, *Tibtech*, 8, 1, 1990.

116. Georges, F., M. Saleem, and A. J. Cutler, Design and cloning of a synthetic gene for the flounder antifreeze protein and its expression in plant cells, *Gene*, 91, 159, 1990.

117. Hightower, R., C. Baden, E. Penzes, P. Lund, and P. Dunsmuir, Expression of antifreeze protein in transgenic plants, *Plant Mol. Biol.*, 17, 1013, 1991.

118. Baertlein, D. A., S. E. Lindow, N. J. Panopoulos, S. P. Lee, T. H. H. Chen, and M. N. Mindrinos, Expression of a bacterial ice nucleation gene in plants, *Plant Physiol.*, 100, 1730, 1992.

119. Tarczynski, M. C., R. G. Jensen, and H. J. Bohnert, Stress protection of transgenic tobacco by production of the osmolyte mannitol, *Science*, 259, 508, 1993.

120. Danheiser, S. L., Quadrant, Calgene form joint food company, *Genet. Eng. News*, 13, 1, 1993.

121. King, R. B., G. M. Long, and J. K. Sheldon, *Practical Environmental Bioremediation*, Lewis Publishers, Boca Raton, FL, 1992, 149.

122. U.S. Environmental Protection Agency (USEPA), Bioremediation of Hazardous Wastes, 600/R-92/126, Office of Research and Development, USEPA, Washington, D.C., 1992.

123. Baker, A. J. M. and R. R. Brooks, Terrestrial higher plants which hyperaccumulate metal elements - a review of their distribution, ecology and phytochemistry, *Biorecovery*, 1, 81, 1989.

124. Steffens, J. C., The heavy metal binding peptides of plants, *Annu. Rev. Plant Physiol. Plant Mol. Biol.*, 41, 533, 1990.

125. Bell, R. M., Higher Plant Accumulation of Organic Pollutants from Soils, EPA/600/SR-92/138, U.S. Environmental Protection Agency, Washington, D.C., 1992.

126. Misra, S. and L. Gedamu, Heavy metal tolerant transgenic *Brassica napus* and *Nicotiana tabacum* L., *Theor. Appl. Genet.*, 78, 161, 1989.

127. Maiti, I. B., G. J. Wagner, R. Yeargan, and A. G. Hunt, Inheritance and expression of the mouse metallothionein gene in tobacco, *Plant Physiol.*, 91, 1020, 1989.

128. Ortiz, D. F., L. Kreppel, D. M. Speiser, G. Sheel, G. McDonald, and D. W. Ow, Heavy metal tolerance in fission yeast requires an ATP binding cassette-type vacuolar membrane transporter, *EMBO J.*, 11, 3491, 1992.

129. Meyer, P., I. Heidmann, G. Forkmann, and H. Saedler, A new petunia flower color generated by transformation of a mutant with a maize gene, *Nature*, 330, 667, 1987.

130. MacKenzie, D., Jumping genes confound German scientists, *New Sci.*, 128, 18, 1990.

131. Lloyd, A. M., V. Walbot, and R. W. Davis, *Arabidopsis* and *Nicotiana* anthocyanin production activated by maize regulators R and C1, *Science*, 258, 1773, 1992.

132. Woodson, W. R., Biotechnology of floricultural crops, *HortScience*, 26, 1029, 1991.

133. Scorza, R., Gene transfer for the genetic improvement of perennial fruit and nut crops, *HortScience*, 26, 1033, 1991.

134. Strauss, S. H., G. T. Howe, and B. Goldfarb, Prospects for genetic engineering of insect resistance in forest trees, *For. Ecol. Manage.*, 43, 181, 1991.

135. Michler, C. H., Biotechnology of woody environmental crops, *HortScience*, 26, 1042, 1991.

136. Knauf, V. C., The application of genetic engineering to oilseed crops, *Trends Biotechnol.*, 5, 40, 1987.

137. Kridl, J. C., D. W. McCarter, R. E. Rose, D. E. Sherer, D. S. Knutzon, S. E. Radke, and V. C. Knauf, Isolation and characterization of an expressed napin gene from *Brassica rapa*, *Seed Sci. Res.*, 1, 209, 1991.

138. Knutzon, D. S., G. A. Thompson, S. E. Radke, W. B. Johnson, V. C. Knauf, and J. C. Kridl, Modification of *Brassica* seed oil by antisense expression of a stearoyl-acyl carrier protein desaturase gene, *Proc. Natl. Acad. Sci. U.S.A.*, 89, 2624, 1992.

139. Voelker, T., A. Worrell, L. Anderson, J. Bleibaum, C. Fan, D. Hawkins, S. Radke, and H. M. Davies, Fatty acid biosynthesis redirected to medium chains in transgenic oilseed plants, *Science*, 257, 72, 1992.

140. Poirier, Y., D. E. Dennis, K. Klomparens, and C. Sumerville, Polyhydroxybutyrate, a biodegradable thermoplastic produced in transgenic plants, *Science*, 256, 520, 1992.

141. Zhu, Z., X. Li, and Y. R. Sun, Expression of human alpha-interferon gene in transgenic rice plant, *Physiol. Plant.*, 82, A31, 1991.

142. Vandekerckhove, J., J. VanDamme, M. Lijsebettens, J. Botterman, M. DeBlock, M. Vanderviele, A. DeClerce, J. Lemans, M. Montagu, and E. Krebbers, Enkephalins produced in transgenic plants using modified 2S seed storage proteins, *Bio/Technology*, 7, 929, 1989.

143. Saito, K., M. Yamazaki, A. Kawaguchi, and I. Murakashi, Metabolism of solanaceous alkaloids in transgenic plant teratomas integrated with genetically engineered genes, *Tetrahedron*, 47, 5955, 1991.

144. Walton, N. J., A fine chemical harvest-plant secondary metabolite production by hairy root culture, cell culture or organ culture optionally after metabolic engineering or pharmaceutical conversion using plant cell biocatalyst, *Chem. Brit.*, 28, 525, 1992.

145. Hiatt, A., R. Cafferkey, and K. Bowdish, Production of antibodies in transgenic plants, *Nature*, 3442, 76, 1989.

146. Benvenuto, E., R. J. Ordas, R. Tavazza, G. Ancora, S. Biocca, A. Cottanea, and P. Galeffi, "Phytoantibodies": a general vector for the expression of hemoglobin domains in transgenic plants, *Plant Mol. Biol.*, 17, 865, 1991.

147. Owen, M. R. L., A. Gandecha, B. Cockburn, and G. C. Whitelom, The expression of antibodies in plants, *Chem. Ind. (London)*, 11, 406, 1992.

148. Proceedings of the Conference on Pesticidal Transgenic Plants: Product Development, Risk Assessment and Data Needs, EPA/21T-1024, Office of Pesticide Programs, U.S. Environmental Protection Agency, Washington, D.C., 1989.

149. U.S. Department of Agriculture, Science and Education and Cooperative State Research Service and Clemson University, Proceedings of the First International Symposium on the Biosafety Results of Genetically Modified Plants and Microorganisms, held November 27–30, Kiawah Island, S.C. Eds. D. R. McKenzie and S. C. Henry, Agricultural Research Institute, Bethesda, MD, 1991, 303.

150. The OECD Workshop on Methods for Monitoring Organisms in the Environment, September 14–17, Ottawa, Canada, Organization for Economic Co-operation and Development, Ottawa, Ontario, Canada, 1992, 73.

151. Bing, D. J., R. K. Downey, and G. F. W. Rakow, Potential of gene transfer among oilseed brassica and their weedy relatives, in Proceedings of the 8th International Rapeseed Congress, Saskatoon, Saskatchewan, Canada, 1991, 1022.

152. Jenkins, J., W. Parrot, J. McCarty, K. Barton, and P. Umbeck, Field tests of transgenic cottons containing a *Bacillus thuringiensis* gene, *Tech. Bull., Miss. Agric. For. Exp. Stn.*, 174, 1991.

153. Lefol, E., V. Danielou, H. Damnay, M. C. Kerlan, P. Valee, A. Chevre, and M. Renard, Escape of engineered genes from rapeseed to wild Brassiceae, *Proc. Brighton Crop Protection Conf. Weeds*, 3, 1049, 1991.

154. Darmency, H. and M. Renard, Efficiency of safety procedures in experiments with transgenic oilseed rape, in *Proceedings of the 2nd International Symposium on the Biosafety Results of Field Tests of Genetically Modified Plants and Microorganisms*, May 11–14, Goslar, Germany, Casper, R. and Landsmann, J. Eds., Biologische Bundesanstalt für Land- und Forstwirtschaft, Braunschweig, Germany, 1992, 54.

155. Downey, R. K., Biosafety of transgenic oilseed brassica species, in *Proceedings of the 2nd International Symposium on the Biosafety Results of Field Tests of Genetically Modified Plants and Microorganisms*, May 11–13, Goslar, Germany, Casper, R. and Landsmann, J. Eds., Biologische Bundesanstalt für Land- und Forstwirtschaft, Braunschweig, Germany, 1992, 17.

156. Dale, P. J., H. C. McPartlan, R. Parkinson, G. R. MacKay, and J. A. Scheffler, Gene dispersal from transgenic crops by pollen, in *Proceedings of the 2nd International Symposium on the Biosafety Results of Field Tests of Genetically Modified Plants and Microorganisms, May 11–14*, Goslar, Germany, Casper, R. and Landsmann, J. Eds., Biologische Bundesanstalt für Land- und Forstwirtschaft, Braunschweig, Germany, 1992, 73.

157. Umbeck, P., G. Johnson, K. A. Barton, and W. F. Swain, Genetically transformed cotton (*Gossypium hirsutum* L.) plants, *Bio/Technology*, 5, 264, 1987.

158. Casper, R. and J. Landsman, Eds., *Proceedings of Second International Symposium on Biosafety Results of Field Tests of Genetically Modified Plants and Microorganisms, May 11–14*, Goslar, Germany, Biologisch Bundesanstalt für Land- und Forstwirtschaft, Braunschweig, Germany, 1992, 269.

159. Hoffman, T., C. Golz, and O. Schieder, Preliminary evidence for horizontal gene transfer between higher plants and *Aspergillus niger*, in *Proceedings of the 2nd International Symposium on the Biosafety Results of Field Tests of Genetically Modified Plants and Microorganisms, May 11–14*, Goslar, Germany, Casper, R. and Landsmann, J. Eds., Biologische Bundesanstalt für Land- und Forstwirtschaft, Braunschweig, Germany, 1992, 247.

160. Palm, C. J., K. Donegan, D. Harris, and R. J. Seidler, Quantification in soil of *Bacillus thuringiensis* var. *kurstaki* δ-endotoxin from transgenic plants, *Molec. Ecol.*, 3, 145, 1994.

161. de Zoeten, G. A., Risk assessment: do we let history repeat itself?, *Am. Phytopathol. Soc. Monogr.*, 81, 585, 1991.

162. *Molecular Ecology*, 3, 1, (Symposium Issue, Ecological Implications of Transgenic Plant Release), 1994, 90.

163. White, J., Personal communication, USDA-APHIS, 1993.

164. Rissler, J., Personal communication, National Wildlife Federation, 1993.

165. Chasseray, E. and G. Duesing, Field trials of transgenic plants: an overview, *AGRO Food Ind. Hi-Tech*, 1992.

166. Giddings, L. V., Personal communication, USDA-APHIS, 1993.

167. Duke, L., Personal communication, Agriculture Canada, 1993.

168. USDA-APHIS, Genetically engineered organisms and products; notification procedures for the introduction of certain regulated articles; and petition for non-registered status, *Fed. Regist.*, 58, 17044, 1993.

169. Stuber, C. W., G. H. Herchel, and D. E. Kissel, Eds., Intellectual property rights associated with plants, in *ASA Special Publication Number 52*, Crop Science Society of America, American Society of Agronomy, Soil Science Society of America, Madison, WI, 1989, 206.

170. Hoban, T. J. and P. A. Kendall, Consumer Attitudes about the Use of Biotechnology in Agriculture and Food Production, USDA-ARS Extension Service, University of North Carolina, Chapel Hill, 1992.

171. Eisner, T. and I. Chapela, Conservation: should drug companies share in the costs?, *Science*, 259, 294, 1993.

172. Juma, C., *The Gene Hunters — Biotechnology and the Scramble for Seeds*, Princeton University Press, Princeton, NJ, 1989, 288.

INDEX